Praise for *Artificially Intelligent*

"With impressive dexterity, David Eliot negotiates the twists and turns in the evolution of AI that have landed us in a world of 'ubiquitous computing,' where we've become cyborgs of a sort in a surveillance state. He invites us to see the 'digital enclosures' created by the likes of Google and Facebook, and warns us that the control of data is power. But things are not hopeless. Eliot insists that if we understand the impact of AI on our lives, and not just passively use it, we can resist the wealth inequality it has created. He cautions us that a storm is coming, and we need to prepare. An important book."

Rosemary Sullivan, author of *Stalin's Daughter* and the *New York Times* bestseller *The Betrayal of Anne Frank*

"*Artificially Intelligent* is a compelling and timely reflection on one of the most transformative forces of our time. David Eliot brings both clarity and depth to a topic too often shrouded in hype or fear. As a professor of marketing and chair of major boards, I see daily how AI is reshaping decision-making, creativity, and governance. This book reminds us that the story of AI is still being written – and that we all have a role to play in shaping its future."

Johanne Brunet, PhD, Full Professor of Marketing, HEC Montréal, and Chair of the Board, SAQ, SQDC, and Groupe TCJ Inc.

"*Artificially Intelligent* astutely identifies artificial intelligence as the Rorschach test of our time: both savior and saboteur to society. In an accessible writing style, David Eliot cogently presents the promise and threats of artificial intelligence. Equally important, he persuasively argues that humans, on an individual and societal level, must take an active role in shaping the evolution of this technology."

Albert Yoon, Michael J. Trebilcock Chair in Law and Economics, Faculty of Law, University of Toronto

"This engaging and hopeful book offers a culturally and historically rich account of AI's human foundations, weaving distant episodes into a cohesive narrative that remains insightful even to well-informed readers. With keen sensitivity to social implications, it illuminates how AI is profoundly reshaping society."

Zsuzsanna Szegedy-Maszak, Curator of Photography, Hungarian National Museum

"Conventional discussions of AI are often trapped in simplistic dichotomies: humans vs. machines, hope vs. fear, capable vs. incapable, before vs. after, and so on. In *Artificially Intelligent*, Eliot offers a social understanding of AI's emergence that frees readers from these limiting frames. He crafts an empowering narrative and invites us to take an active role in this unfolding process."

Jean-Frédéric Morin, Professor of Political Science, Université Laval

"*Artificially Intelligent* is the most lucid and approachable discussion of artificial intelligence I have read. Throughout, Eliot intertwines the story of his own interest in AI with stories of some of the pioneers of the field, from the eighth-century figure who gave us the word 'algorithm' to the leading figures of today. He is clearly excited at the possibilities for AI to improve our lives but argues that the way it is being implemented politically is leading us in the wrong direction. AI is not some vision for the future but rather is already deeply embedded in many systems that shape our lives today. Eliot explains the trade-offs we are making, explores the potential negative impacts, but nevertheless ends with hope. This is a book that I would recommend to anyone seeking to understand the larger context of the adoption of AI."

Virginia Haufler, Professor, Department of Government and Politics, University of Maryland, College Park

Artificially Intelligent

The Very Human Story of AI

David Eliot

 UNIVERSITY OF TORONTO PRESS

Aevo UTP
An imprint of University of Toronto Press
Toronto Buffalo London
utppublishing.com

Library and Archives Canada Cataloguing in Publication
Title: Artificially intelligent : the very human story of AI / David Eliot.
Names: Eliot, David, author.
Description: Includes bibliographical references and index.
Identifiers: Canadiana (print) 20250228785 | Canadiana (ebook) 20250228815 |
ISBN 9781487567675 (cloth) | ISBN 9781487569006 (EPUB) |
ISBN 9781487568986 (PDF)
Subjects: LCSH: Artificial intelligence. | LCSH: Artificial intelligence – Social aspects.
Classification: LCC Q335 .E45 2025 | DDC 006.3 – dc23

ISBN 978-1-4875-6767-5 (cloth)   ISBN 978-1-4875-6900-6 (EPUB)
ISBN 978-1-4875-6898-6 (PDF)

Printed in the USA

Cover design: Alex Camlin
Cover image: Da Vinci, Leonardo. Mona Lisa. 1503–1506. Oil on wood.
Louvre, Paris. Wikimedia Commons.

Excerpt from *The Fellowship of the Ring* by J.R.R. Tolkien reprinted by permission of
HarperCollins Publishers Ltd. © 1954 J.R.R. Tolkien.

We wish to acknowledge the land on which the University of Toronto Press operates. This
land is the traditional territory of the Wendat, the Anishnaabeg, the Haudenosaunee, the
Métis, and the Mississaugas of the Credit First Nation.

University of Toronto Press acknowledges the financial support of the Government of
Canada, the Canada Council for the Arts, and the Ontario Arts Council, an agency of the
Government of Ontario, for its publishing activities.

 **Canada Council
for the Arts**   **Conseil des Arts
du Canada**    **ONTARIO ARTS COUNCIL
CONSEIL DES ARTS DE L'ONTARIO**
an Ontario government agency
un organisme du gouvernement de l'Ontario

Funded by the   Financé par le    Canada
Government   gouvernement
of Canada   du Canada

For my great-aunt Sarah,
who gifted me the book that started this journey.

Contents

ACT 3: FRANKENSTEIN'S MONSTER

Acknowledgments

Although my name is on the cover, this book would have been impossible without the help, teachings, support, and labor of countless individuals. Although I cannot mention everyone, there are a few names that deserve a special thanks:

- My supervisors (past and present), Rod Bantjes and David Murakami Wood.
- Daniel Quinlan, for believing in this project and taking a chance on me.
- Hugh Barker, for editing my first draft.
- Kate Laumann Wallace, who painstakingly fact-checked the book and served as a historical consultant.
- Every professor in the St. Francis Xavier University Sociology Department.
- And my parents, for simply being my parents.

Prologue

I can't help but feel romantic about the idea of academia: the tireless pursuit of knowledge and the desire to create a better world. When I was younger, I counted the days until my next visit to my grandparents' house. A classics professor and an archaeologist, they were a living time capsule of the golden years of the university. The walls of their home were disguised by oak bookshelves, artwork, and antique clocks. Their parties were populated by the most intriguing and quirky of characters. They were like something out of a movie.

I was ten years old when my grandfather passed. I never had the opportunity to get to know him as a man instead of an image. My grandmother passed during my freshman year of university, just as I took my first steps into the world of academia. I often wonder what they would think of the ivory tower I stumbled into …

When I meet people, and they discover that I am an AI researcher, the first questions they ask are almost always the same: "Are you scared? Should I be scared?" We've all heard the stories and had the barroom conversations. There's no point in beating around the bush. The conversation about AI can be terrifying. From the far-fetched fantasy of killer robots taking over to the very real threat of mass automation, it seems like AI captured the public imagination almost overnight.

Governments are hastily drafting new AI legislation. World leaders are convening special advisory groups, composed of every major tech CEO, to help them understand the threats that AI presents. Simultaneously, companies are automating their workforces, and policing agencies are implementing untested and unreliable AI systems. It is the Wild West.

But if you want the truth, I am not scared. I am hopeful.

I wholeheartedly believe that AI has the potential to be the most liberating technological advancement in human history. It certainly unlocks more human potential than any innovation since the steam engine.

There is definitely a lot that keeps me up at night: things I have seen, economic models I have studied, and theories I have helped write. I am not scared of the technology itself. But I do fear the path we are following – and I partly blame my own kind, academics, for it.

The truth is that, like the steam engine before it, the emergence of practical AI is shifting the foundations of our society. I'm sure you already know this, and already feel it, even if you don't know how to describe it. On its own, this shift is nothing to be afraid of. Personally, I find it exciting to live through such a key moment in human history. The real question is, what kind of society will we have when the dust settles? What happens when the new foundations are built?

Rome wasn't built in a day, and massive societal changes don't occur fast or with a single law. We are at the beginning of a slow process – a series of events and decisions that will shape the world we inhabit and the one we get to leave behind. Many decisions have already been made. More are in the pipeline. My fear is that the world we are making is not being made democratically. The decisions about AI are not being made democratically. There is no grand conspiracy, no secret cabal pulling the strings from behind the stage. Our problem is much more mundane.

To meaningfully participate in democracy, one must have knowledge. You must be able to understand the reality of your situation to be able to advocate for yourself and to ask for what is best for you. But AI and the problems we are facing are complex. Understanding the technology itself requires a degree, not to mention comprehending the sophisticated social and legal systems that it collides with. This complexity is why governments have gathered tech CEOs to help them understand and address the problems. The issue here is that it is now the CEOs who are getting to dictate how we understand the issues.

If we understand our condition, we can ask for the world we wish to see. It is not my intention in this book to pitch my politics to you. I will be completely satisfied with the outcome if I just help someone understand AI beyond the headlines and give them the ability to meaningfully speak for themselves.

At the end of the day, I actually think that the problems we face are quite simple. The technology is complex but understandable. You do not need to understand it at a PhD level to be able to begin to understand the realities we are facing together as a society.

I am writing this book because the purpose of academia should not just be to produce knowledge but also to share it. What use is producing knowledge if we cannot effectively share it with those who need it most? I want to share with you the complexities of AI and the social systems intertwined with its rise: the politics, the battles, the greed, and the innovation. I wish to go beyond the surface level to reveal the roots of the moment where we are, the decisions and actions that got us here. By understanding the roots of our shared present, we can learn to better understand our own unique experiences. I hope that, once we dig up these roots, you can see the complex systems in your own life and understand that the events you are witnessing are interconnected to a system bigger than is humanly imaginable.

In academia, we write for each other. We often write big words to feel important and accepted. Knowledge must be shared through accepted formats that you often need formal training to interpret. We often do so for good reason, but in doing so, we actively cut you out of the conversation. I am not writing this book as an academic. Everything I write is grounded in fact. I am rigorous. But this book is not just for academics. It is for anyone who wants to understand how we got here. As such, I will tell the story of how we got here as it is best understood: as a story.

So, as with most great stories, we start from the beginning, with a war, an inheritance, and the daughter of one of the world's greatest poets.

ACT 1

The Foundations

War, Inheritance, and the Poet's Daughter

CHAPTER 1

The Household Name That No One Knows

Before I was an academic, I was a magician. Not the top hat–wearing, rabbit-out-of-a-hat type of a magician though. There were no big boxes and no dancing ladies. All the audience saw on stage was me, a small table, and a deck of cards. I had one job every night: do the impossible and don't get caught cheating.

The thankless thing about doing magic is that, when done right, your audience will never know how hard you had to work to pull it off. In reality, the audience only ever sees about 15 percent of the show. The other 85 percent is hidden from them. But ask any magician, and they will tell you that that hidden 85 percent is where the real magic happens: the hidden moves that we spend lifetimes perfecting; the vital pieces of information we withhold from the audience to create the illusion of the impossible.

You could probably go your whole life without encountering the underworld of magic: networks of private clubs, companies, and conventions; a community of people dedicated to building and sharing secrets. Meetings between magicians are fascinating because we all have stories about tricks gone wrong or new sleight-of-hand discoveries. But we can never tell anyone else about them.

It's not because of some secret or solemn oath we take, but because once others know how our tricks work, the illusion is shattered, and we lose our power.

Magic shows are at their best when the audience is oblivious to how the tricks are done. If I fool you with a trick, you may think about it for the rest of your life. But if you go home and find out how it works on YouTube, you might never think of it again. It's the mystery that gives us power.

AI is a lot like magic. The first time you see a good AI system, it really does look like magic. Yet it goes deeper than that. AI often has power over us because we don't understand what is going on behind the curtain. But pull back the curtain, reveal what's going on behind the scenes, and it's actually quite mundane. Quite understandable. Pull back the curtain, and it loses its power. Unlike magic, I don't think the secrets of AI should be gatekept. They belong to everyone, and everybody has the right to understand how AI works.

I decided to quit magic during the third year of my undergraduate program. At the time, I was spending three months each summer touring full time, and the lifestyle was starting to wear on me. Fortunately, my university offered me a job. It was my first time working at a university. They gave me a small grant to do some research on Canadian politics. A paid three-month contract was an extraordinary opportunity for an academic my age. They gave me a little office and expected me to work eight hours a day on my project. They were fools.

Although I did produce the report I had promised, I spent most of my days reading about something else, a new technology: artificial intelligence. About a month before my contract started, I read an article about a new computer program that could write like a human. One researcher said it could even read and answer questions. I emailed everyone I knew who worked in the industry, and they all confirmed that it was real. Eventually, I found samples of the text produced by the program. They were mesmerizing. The implications hit me like an out-of-control semi-truck traveling at 100 miles per hour on the freeway.

The year was 2019, and the system was GPT-2.

I spent much of the summer in my asbestos-lined basement office researching AI. I reached out to experts when I had questions and wrestled with computer science concepts, both basic and advanced. I didn't know it at the time, but I was building the foundations for the rest of my academic journey.

Part of the way through the summer, I got a phone call from a small theater festival in Saskatoon. They had had an act drop out and needed a quick replacement. The money would be good, probably more than the university was paying me for the full summer, and it was only two weeks; no one would even realize I was gone. So I said yes.

One day, after a show, I sat at a little greasy spoon diner with a legendary British poet. He was one of those people that you could listen to for hours. And if you let him go, he certainly would talk for hours.

We exchanged pleasantries, and he showered me with stories of his recent adventures. Eventually, when my time to talk emerged, I told him about my fascination with AI. As I finished my rambling, he looked at me with eyes that foreshadowed a spectacular story.

"Tell me David," he said with a grin. "Have you ever heard the story of how algorithms got their name?"

THE INHERITANCE

Muḥammad ibn Mūsā al-Khwārizmī might have one of the most fascinating legacies of any historical figure. He is the household name that no one has ever heard of, a name both familiar and shrouded in mystery. Even those familiar with his legacy know shockingly little about him.

Born in the year 780, he is widely believed to be from the Middle Eastern region of Khwarazm, a beautiful oasis sandwiched between vast lifeless deserts.[1] Others speculate that his roots lay thousands of miles away in the town of Quṭrabbul. Without definitive records, the truth eludes us.[2] Regardless of where he was from, we know with certainty that his life journey led him to the newly erected city of Baghdad.[3] It was here that he would indelibly inscribe his name into the story of AI.

The city would have been a magnificent sight for al-Khwārizmī to behold. Less than twenty years before he had been born, the caliph (king) of the Islamic Empire, al-Manṣūr, had decided to move the seat of his reign to Baghdad.[4] There was only one problem: at the time, there was no such thing as Baghdad. All that existed was sand, bush, and a river. But the caliph dreamed of a city that would usher in a new dawn for the fledgling empire.

The Islamic Empire had undergone something of an intellectual dark age, and the caliph believed that the new city would stand as a beacon, attracting the greatest minds of the empire to one location. He imagined a collision of culture and mind, art and philosophy. Al-Manṣūr died before seeing his vision blossom, but through the construction of Baghdad, he laid the groundwork for the Islamic golden age.[5]

Al-Khwārizmī arrived in Baghdad during the early days of this golden age, a period of unparalleled cultural growth and scientific discovery. The city had quickly become the largest metropolitan center in the world, and formidable intellectual institutions had been swiftly established.[6] Thousands of the brightest minds in the world flocked to Baghdad in hopes of being part of the history that was being written. Although it is impossible to know their motivations, it is all but certain that many of them were not satisfied with merely being a part of history; they wanted to be the ones to make it. But for all the great thinkers who walked through the city's gates, few became immortal. Our story is about one of the lucky few.

A man of many talents, al-Khwārizmī probably had no issue in Baghdad with fitting in. He worked in the House of Wisdom, one of the city's most coveted academic institutions.[7] Here he established himself as a true master of the mind. He was a gifted translator, historian, scientist, and astronomer, but beyond all else he was a mathematician. However, the math he did would have looked very different from the math we do today.

It is strange to think of math as an invention. It's often taught to us in primary school as something that simply exists for us to memorize, practice, and, hopefully, master. It is depicted as something that just exists and is true. New discoveries in mathematics often feel distant and abstract. But mathematics itself is a human invention. It is the action of describing the natural patterns of the world around us. To do so, we invent tools that allow us to share our observations with others. These tools include numbers. Can you imagine a time before the invention of numbers?

To be fair, we have been using numbers for tens of thousands of years, the earliest appearing in the form of tally marks. But the system that we refer to as numbers today was not invented until around the sixth or seventh century AD by Indian mathematicians.[8] It had not yet been introduced in Baghdad upon al-Khwārizmī's arrival. We know that because al-Khwārizmī would eventually be the one who introduced the Indian numeral system to Baghdad and is credited for spreading it throughout the Middle East and Europe.[9]

Today, mathematicians invent new mathematical methods to solve complex and abstract problems. By contrast, the mathematics of ancient Baghdad often solved very personal ones. Mathematicians were called upon to recognize patterns that could meaningfully change the daily lives of citizens. By recognizing and giving meaning to these patterns, one could manipulate the world around them.

A significant problem that hung over the head of every mortal man in the Islamic Empire was that of inheritance.

The Qur'an, Islam's holy book, describes complicated rules that must be followed when dividing a man's wealth after his death.[10] These rules were, and still are, highly important to the Islamic faith and, by extension, Islamic society.[11] The rules exist as a series of intertwined rights and restrictions. Different degrees of relatives, and other community members, are entitled to vastly different amounts. They are all affected by the number and status of other beneficiaries. The division of inheritance has massive social and religious importance in Islamic society. So the pressure to get the calculation correct is immense.

Nowadays, a Muslim nearing death can simply Google "Islamic Inheritance Calculator," fill out some variables, and receive a precise breakdown of how their money and property should be distributed. But remember, at the time of al-Khwārizmī, not only were there no calculators, but there were no written numbers with which to create equations! The task of dividing one's inheritance was daunting and stressful. Luckily for the inhabitants of the Islamic Empire, al-Khwārizmī sought to change that.[12]

With direct encouragement from the leader of the Islamic caliphate, al-Khwārizmī tackled the inheritance problem head on and, in the process, wrote the first note in the symphony of AI that we are listening to now.[13] His book, *The Compendious Book on Calculation by Completion and Balancing,* did not begin with the problem of inheritance itself but rather by developing and describing relationships between numbers and variables.[14] It describes, using only words, how to solve complex theoretical problems such as quadratic equations.[15] The text itself is widely considered to be the founding of the discipline of algebra.

With the theoretical concepts established, al-Khwārizmī applied the method he had imagined to the process of distributing the wealth of the dead.[16] He leveraged the theoretical mathematics he had developed to create in-depth instructions that, when followed, would result in the proper allocation of funds. All someone had to do was follow the instructions he wrote, applying them to their situation, and they could accurately and painlessly distribute their inheritance, needing only to perform simple calculations themselves.

Take a second to think of the inheritance calculator you might find on the internet. If you are not Muslim, you may instead choose to think of a mortgage calculator. Such devices allow you to provide relevant information (the variables); then, at the click of a button, it will run your information through an algorithm, answering your complex financial question. What al-Khwārizmī did was no different, except that a human needed to "crunch"' the numbers instead of a computer. In fact, what al-Khwārizmī had invented was the algorithm.

The word "algorithm" comes from the Latinized pronunciation of his name: *Algoritmi*.[17]

Algorithm ...

It's a word we hear so often that it seems to have lost all meaning. The TikTok algorithm. AI algorithms. The Google News algorithm. But what are they? Why do they matter? And why do they seem to have so much power?

Unfortunately, the answer to some of those questions will need to wait until later. But what an algorithm is can be answered now.

Al-Khwārizmī had a problem. His problem was not his inheritance, as he was smart enough to figure that out on his own. His problem was that others needed to figure out their inheritances. So he developed a set of instructions that they could follow to achieve their goal. The instructions were designed to take into account each user's specific circumstances. That is really all an algorithm is: a set of rigorous instructions that, when followed, will fulfill a desire. All you need to do is fill in the variables and follow the steps.

The Google advertisement algorithm ... a set of instructions that, when followed, given the variables of your personal information, will provide a relevant targeted ad.

The TikTok algorithm ... a set of instructions that, given information on your TikTok history, will tailor your "for you page" so you become addicted to the app and endless scrolling.

The ChatGPT algorithm ... a system that follows a set of instructions to write a text based on a prompt you provide.

But these algorithms take a much different form than the simple inheritance algorithm written by al-Khwārizmī, and their social effects are far wider reaching. These algorithms are so powerful because they are not performed by humans – they are executed by computers.

To understand how algorithms jumped from human hands to machines, we must look in the most unlikely of places. Not in the computer labs of MIT or the garage of a soon-to-be billionaire. No, instead we turn our gaze toward the first computer programmer, the daughter of one of the world's most famous poets – a woman so far ahead of her time that she wrote the first computer program 100 years before the first computer was even built.

CHAPTER 2

The Enchantress of Numbers

I am not a skilled enough writer to describe the British poet Lord Byron. The words that first come to mind are womanizing, revolutionary, wordsmith, and cruel.[1] He is probably best described as he was by a former lover: "mad, bad and dangerous to know."[2] Lord Byron infatuated the British public with his bold antics, just as he did with his poetry. The women of nineteenth-century Europe swooned for him, and men wanted to be him. He was the precursor to the modern rock star.

Unfortunately, as is the fate of many who live with unchecked fame and excess, Byron's actions had consequences. By the time he found himself engulfed in scandal, accused of incest with his sister, he had already made numerous powerful enemies.[3] Apparently sleeping with the wife of every nobleman you meet can have its drawbacks. Facing mounting pressures, both financially and socially, he fled his home country of England, never to return.

During his exile, while he was gallivanting carelessly around Europe, he met and befriended a young poet, Percy Shelley, and his girlfriend, Mary. The three would become inseparable for a time, and one rainy summer night in Geneva, Lord Byron would accidentally inspire one of the greatest stories ever written.[4]

Vacationing with two other friends, the group found themselves victim to what felt like a never-ending rainstorm. As they were trapped inside, Byron suggested that the party wait out the storm by telling ghost stories.[5] The catch – each member of the group was to write their own ghoulish tale. With two world-class poets, one of whom was arguably the greatest living poet of the era, one would expect nothing less than top-notch horror: the kind of tales that would keep even the most level-headed of listeners on edge for weeks. At least one such tale was written that night, but it was neither of the poets who wrote it.[6]

When the group reconvened after their writing sessions, Mary shared with them the most frightening of chronicles. She had written a stark warning of the foolishness of man and the risk of our technological ambitions.

Her story began when two adventurers found a strange, and somewhat deranged, man in the Arctic. The man had once been a great scientist. But his moral compass had been clouded by his own ambitions and belief in his intellectual superiority. Finding himself surrounded by death, the scientist took it upon himself to become God. He decided to attempt the unthinkable and become the master of life itself, creating new life in his laboratory. He robbed graves to find body parts and sewed them together to construct his perfect specimen. But when the scientist succeeded, when his years of study and hard work finally paid off and his creation came to life, he looked upon it not with pride but with horror.

By the time the adventurers found Frankenstein in the Arctic, he had fully come to grips with the monstrosity he had created. He had sworn an oath to pursue the monster to the end of the earth and to rid the world of his error. He was in the Arctic sleeplessly hunting the beast.

Surprisingly, although he commits numerous evil acts, the monster himself is a sympathetic character. He did not ask to be created, and he is scared, alone, and outcast by the world that he finds himself in. Although he murders, it is only in reaction to the pain and loneliness he is made to feel by the rejection of his creator. It was the cruelty of man that turned him into a monster, not his fate. In the end, the reader is left to ask who the true monster is: the man or his creation.

Eight years after first hearing the story of Frankenstein, Lord Byron died.[7] It was not at the hands of a jealous lover or vengeful husband as many expected. Instead, his life ended in a fittingly dramatic way for a romantic such as him: fighting alongside the Greeks in their revolution against the Ottoman Empire.[8] Although Byron probably would have preferred to have died by the sword or in artillery fire, or perhaps not at all, it was illness that eventually took him. He left behind a young daughter whom he barely knew, Ada. When Mary Shelley read *Frankenstein* to the group, she had no way of knowing that the seductive British poet sitting across from her had fathered a child whose mastery of science would rival that of Dr. Frankenstein himself and who would lay the groundwork for the making of a real monster.

THE POET'S DAUGHTER

Ada Lovelace grew up relatively sheltered from her father's influence. Her mother, Annabella Milbanke, made sure of that. Annabella was by all accounts an extremely rational, kind, and level-headed woman – the polar opposite of her husband.[9] Where Byron reveled in sin and gave in to the temptations of excess, Annabella was modest and devoutly religious, with strict morals. What brought the two together was their brilliant minds.

Annabella was exceptionally gifted, and due to her family's deep pockets and social status she received a prestigious private education.[10] She quickly established herself as a prodigy in mathematics and science.[11] Although Lord Byron could have courted any woman in England at the time, he fell deeply in love with the intellect of Annabella.[12]

Initially hesitant of Byron, Annabella found herself in his presence on many occasions. Although he disgusted her at times, she was drawn in by his mind. As her infatuation with the poet grew, her more sensible side remained acutely aware of his shortcomings. She was not blinded by Byron's charm or his brilliance, and understood that he was a deeply troubled man with dangerously reckless tendencies. Her one miscalculation was that she thought she could change him. She hoped she could provide the love and structure he needed to overcome his darkness and redeem his soul.

The two married in the winter of 1815.[13] At the time, public opinion of Byron had soured. The realities of his personal life had merged with rumors, and many believed that he had married Annabella in an attempt to rehabilitate his image.

Shortly after their marriage, Annabella became pregnant with their daughter.[14] Sadly, the union began to sour at the same time. Lord Byron's close relationship with his half-sister became increasingly suspicious to Annabella, and she began to believe that it had become incestuous in nature.[15] Additionally, problems with Lord Byron's personal finances drove him to drink excessively.[16] Annabella was often the recipient of his drunken rage.[17] Byron's behavior became erratic, and he became increasingly dangerous to be around.

Annabella was convinced that her husband had gone mad and tried tirelessly to help him in his recovery.[18] She enlisted the help of medical experts, did her own research, and pleaded with him to change his ways. But her efforts were futile. In 1816, Annabella left with young Ada to join her parents for what Lord Byron thought was a temporary visit.[19] He never reunited with his wife or child.[20]

Although Annabella tried from a distance to reason with her husband, the realities of the situation quickly became apparent to her. To Lord Byron's great disbelief, she requested a separation. It was soon after, in a heartbroken state, that the poet fled the country and began the journey that would lead him to Mary Shelley and his death in Greece.

Young Ada, to her mother's chagrin, retained many of her father's attributes. She was bold and adventurous, and she had a disdain for the rules of society. Annabella tried to temper her late husband's traits, forcing her child to endure intense exercises of self-discipline and persuading her away from the pursuit of literature.[21] To replace literary education, Annabella made sure her daughter received the best and most interesting education in mathematics and the sciences that her family's fortunes could provide. Like her mother, Ada proved to possess a gifted mind for mathematics and quickly excelled.[22]

While still a teenager, the gifted young Ada received an invitation to a party hosted by the legendary inventor Charles Babbage.[23] Her acceptance set the stage for a meeting of minds that sparked the match of modern computing.

Babbage was a small-framed man who stood large in stature. Like many great minds, he understood his place in history long before anybody else did. When he first arrived to study mathematics at Cambridge,[24] he found himself disappointed by his professors and the other students. Before arriving, he had been reading the newest, most exciting publications in the field. One can only imagine that his mind must have been overwhelmed and endlessly occupied by the possibilities and excitement of reading work at the edge of human knowledge. But his ambition outpaced the educational realities. He found himself stuck in stuffy lecture halls, relearning the fundamentals of mathematics. The old and known must have seemed so mundane when Babbage had the ability to dream of the future.

The dull instructions of his classes gave Babbage the impression that he was being trained to be a computer and not a free thinker. At the time, the word "computer" meant something very different from what it does today. Instead of referring to a device, "computer" referred to a person. A computer's job was to perform calculations. If an engineer was building a bridge, the computer would execute the mathematical formulas necessary to ensure the bridge would not collapse. Basically, any activity or job you use a calculator for today required a human computer back then. It was the most tedious job a mathematician could have.

During the nineteenth century, a common job for computers was the construction of mathematical tables. These tables were gigantic books that provided the answers to complex mathematical formulas commonly used in equations. Each line of the table would provide the formula's answer when slightly different variables were plugged in. These tables were vital for advanced mathematics as they saved countless hours of calculations. The problem was that human error often crept into the tedious job of calculation, and the tables that mathematicians trusted sometimes provided wrong answers.

One day, in a dream-like haze, Babbage was sitting at a campus bar staring deep into the pages of a calculation table.[25] In his mundane mathematical prison, he began to see glimmers of the future. Beautiful and simplistic patterns started to jump out at him. The sight of Babbage staring aimlessly into a book of mathematical tables proved too intriguing for his friends. They interrupted his trance to ask: "Well Babbage – what are you dreaming?"

"I'm thinking," he responded, "that all of these tables may be calculated by a machine."[26]

With those words, the journey to the modern computer began.

When Babbage and Ada met, nearly twenty years later, he was no longer a student but a well-established mathematician, with a side hobby designing the world's first computers.[27] At the time, he held the position of Lucasian Professor of Mathematics at Cambridge, one of the most prestigious academic positions in the world.[28] Just 150 years before him, the position had been held by Sir Isaac Newton. And 150 years later, it would be filled by Stephen Hawking.[29] Needless to say, it was a big deal.

After dreaming up a mechanical computer, Babbage had received a large government grant to pursue his vision.[30] His original design, the Difference Engine, was to be a steam-powered machine that could solve specific sets of problems and mechanize the production of mathematical tables with blistering accuracy.[31] Unfortunately, due to disputes with his engineering team, and a mismatch between vision and funding, Babbage never managed to complete a fully functional Difference Engine.[32] He did, however, produce a working prototype as proof of concept for his funders.[33] Instead of being powered by steam, the prototype was operated by a hand crank. And instead of being able to calculate a wide range of formulas, it could only do one. Although simplistic when compared to today's standards, or Babbage's dreams, the prototype was a monumental leap forward for technology.

When Babbage met Ada in 1833, he was mesmerized by the seventeen-year-old's mind.[34] Maybe he saw a bit of himself in her: a free thinker who refused to be held back by the expectations of society; a brilliant mind with the ability to comprehend the future.[35] Whatever his reasoning, Babbage made the decision that night to introduce young Ada to his machine.[36] Looking upon the device of steel and mind, Ada peered into the future. She was entranced by its ingenuity and at once understood its significance.[37]

After their first meeting, Babbage and Ada kept in close contact, sharing their minds with one another. Constantly shocked by the adolescent's skill, Babbage gave her the nickname "the enchantress of numbers."[38] But no matter how talented, Ada needed to wait to make her mark on history. In 1835, she married. A year later, she gave birth to her first of three children, a son named Byron.[39] In 1838, her husband was given the British title Earl of Lovelace, replacing Ada's title of "the enchantress of numbers" with that of Lady Lovelace.[40] Mathematics would take the backseat for the time being as Lady Lovelace worked hard to keep her family in order. There is no doubt that the young Byron was not much better behaved in his infancy than his grandfather had been in adulthood. And one may imagine that even a well-behaved Byron is difficult to manage.

By the time Lady Lovelace returned to the field of mathematics, the groundwork had been laid for her to make a momentous impact. Seven years after showing the young Ada his first machine, Babbage revealed his plans for a second.[41] It was to be a much grander, much more ambitious machine: the Analytical Engine.[42]

Much like Ada at Babbage's first party, the Difference Engine had been juvenile – full of potential but limited by a lack of maturity. Babbage's plans for the Analytical Engine had much more in common with the grown Lady Lovelace who stood before him now. It was complex, multifaceted, and ambitious far beyond its years. Where the Difference Engine could only compute a handful of suitable formulas, the steam-powered Analytical Engine was to be a universal computer, with the capability to calculate any mathematical formula.[43] In Ada's absence, Babbage's ambition had conjured up unthinkable innovations. He had drawn the designs for a mechanical system that wouldn't just perform calculations but could also save the results in another section of the machine so that they could be used in future calculations. Babbage had found a way to give his machine memory.

Ada, like Babbage, believed that the new machine would change the world. She felt flattered that the inventor let her into his inner circle and allowed her to provide comments on his drawings and theories. Babbage, for his part, was probably ecstatic that the gifted Lady Lovelace was willing to help him with the design of his device. Most people he told about his vision thought him to be a lunatic. There wasn't a government in the world that would even consider funding the machine's construction. But Babbage knew Lady Lovelace understood: a machine that could flawlessly perform any calculation would revolutionize mathematics in a way that hadn't been seen since the invention of arithmetic by al-Khwārizmī.

It is unclear how long it took for Babbage to discover that Lady Lovelace did not entirely share his vision. In fact, hers was much grander. Lovelace realized that, while Babbage possessed a visionary mind, her friend was thinking too small. She had no plans to merely mechanize mathematics. In the Analytical Engine she saw a way to mechanize everything.

Lady Lovelace believed that the world around her could be completely modeled by numbers. Just as physicists model the behavior of natural phenomena with numbers, she could model the relationship between notes of music.[44] She could represent harmonies with mathematical patterns. And if she could represent music with numbers, then the numbers could be given to the Analytical Engine, as well as an algorithm to understand them. And maybe, utilizing its memory, the engine itself could write beautiful mechanical symphonies. What she realized, and what Babbage had missed, was that his machine could be modified to recognize all sorts of symbols and accept infinite different types of data. With proper programming, it could address and solve an untold number of human problems. Lady Lovelace was the first to recognize that what Babbage had conceived of was not a calculator at all but the world's first general computer.[45] The machine just needed to be told what to do. It needed a way to communicate and to receive algorithms that would instruct its mechanical features in the dance of computation.

To communicate with the machine, Babbage envisioned a punch card system. He borrowed the idea from another transformative machine of the time: the Jacquard Loom.[46] The loom automated the tedious job of making textile products like blankets and rugs. The mechanical arms of the loom replaced the trained hands of weavers, skillfully and tirelessly entwining lengths of fabric to create beautiful tapestries. The ingenious design of the loom made it programmable. A chain of punch cards could be fed into the loom that would direct the movement of the rods. The punch cards were large paper cards, each with holes punched through them – hence the name. Each hole instructed the weaving of a specific thread. Once a set of punch cards was produced for a design, it could be copied and used to program countless looms to make the same design over and over again. Such a system seemed perfect for communicating with the Analytical Engine.

Lady Lovelace was no stranger to the Jacquard Loom or the punch cards used to control it. After first meeting Babbage, and becoming enchanted by his Difference Engine, she had embarked on a trip to the British countryside with her mother to witness the mighty machines of the Industrial Revolution first-hand.[47] As fate would have it, she had taken particular interest in the Jacquard Loom.[48] Later in life, musing on the similarities between the two machines, she wrote, "We may say most aptly that the Analytical Engine weaves algebraical patterns just as the Jacquard Loom weaves flowers and leaves."[49]

Although there is no supporting evidence, I like to imagine that, on one of her many factory visits, the young Ada may have encountered a series of destroyed looms tossed to the side of a factory floor. She might have asked, "What happened to those looms? What mechanical failure must I avoid with my machines to avert such destruction?"

To which a foreman could have responded, "These machines were not destroyed by their failure but by the consequences of their success."

Ever the curious mind, Ada would inquire further: "But who would destroy such a marvelous machine?"

The foreman would have shuddered. "The Luddites, of course – those who fear progress."

She would have learned about the Luddites in one of her lectures on current affairs. They were skilled tradespeople, many of whom worked in the textile industry. They were said to be barbaric mobs who broke and burned machinery designed to make their work more efficient. They feared the progress that modern technology brought and didn't have the ability to recognize that the technology was improving society and making textile goods cheaper and more accessible to the masses.[50] No wonder the foreman shuddered at the thought of them. He was better off working with the more predictable and reliable machines than with dim and moody skilled workers.

Or maybe, with her great wisdom, she saw through the narrative that the Luddites had been painted with and sympathized with their motivations. Her father had been a staunch defender of the Luddites.[51] As he saw it, the Luddites were not simply a gang of thugs – they were a political movement. A complete reading of their history shows that it was not the machines they were protesting. They had no fear of the machines themselves. Instead, they disliked the way the machines were being implemented and the effects the machines had on their communities.

The machines could have been used to inspire more innovation in the textile industry, but the Luddites felt they were being used to stifle it in the name of profit. Skilled workers and artisans, who had trained for years and generated enormous profits for their bosses and foremen, were suddenly being thrown on the street and replaced with low-skill workers, who could be paid a fraction of what the tradespeople had been paid. There was no opportunity for them to repurpose their talents. And the bosses showed little interest in finding ways to use the new technology to make basic textiles more accessible, while creating more time for the skilled workers to innovate where the machines could not.

In contemporary economics, the Luddites are often dismissed with a simple mention of the "Luddite fallacy." This dismissal is based on the theory that, although technical innovation and automation may create unemployment for those who are de-skilled, the long-term economic benefits for society will outweigh the short-term harms. Indeed, over time, the machines caused the prices of textile products to fall, creating higher demand, meaning more workers were needed in the factories than before. Yes, the Luddites lost their jobs, and yes, their families and communities suffered, but more jobs were created elsewhere because of their involuntary sacrifice. But were these economic gains truly good for society at large, and do they not ignore the very real suffering of the Luddites?[52]

Maybe Lady Lovelace didn't think of the Luddites during her visits after all.[53] Yet it is only fitting that the story of computers is so intertwined with the story of textile manufacturing. The road that Lady Lovelace was paving would rekindle the ideas of the Luddites 150 years later and once again raise questions about how much we should ask workers to sacrifice. But that is a story for later.

After re-sharpening her skills in mathematics, Lady Lovelace was asked to complete a relatively simple task. Babbage had given a lecture on the Analytical Engine in Italy, and it had been translated and published in French.[54] Her assigned job was simple: translate the text to English so that it could be published in an English journal. But Lady Lovelace was her father's daughter – a rule bender and a free spirit. Instead of a simple translation, the finished paper included numerous notes and addenda about the Analytical Engine, written and imagined by Lovelace herself.[55] Her additional notes ended up being almost as long as the original paper she had been asked to translate.

The notes included brilliant insights, immortalizing her views on the potential of the Analytical Engine. She pays tribute to the punch cards, giving their integration into the machine credit for its potential. Without the punch cards, the machine would be limited to solving problems only with the values and instructions built into it. But with the punch cards, the machine's function became adaptable. As long as a human problem could be made legible to the machine by the punch cards, the machine could compute it.

Her notes muse about the potential uses of the engine made possible by the punch card system. She describes in depth her theory of how they could be used to algorithmically generate music. But she gives a warning to those who may get too ambitious. She wrote that the machine may be powerful, but it could never be creative. It can only do what we already know how to do.[56] She finished by signing her name on the timeline of human history.

In note G, her final note, she provides a demonstration of how the Analytical Engine could be commanded. She takes a complex equation and breaks it down into simple steps that could be programmed into punch cards to be executed by the Analytical Engine. With each pen stroke she came closer to history, as she wrote a set of instructions, an algorithm, which could be read by a machine. The instructions could then be understood and executed in order to achieve a goal. This algorithm was different from any other algorithm written before it. From al-Khwārizmī onward, every algorithm had been designed to instruct a human being. The human was the one who had to complete the task at hand. But Lady Lovelace's algorithm was to be read by a machine, and the task was to be performed by a machine. *She had written the first computer code and created the first computer program.*

The date of publication was September 1843. Babbage's Analytical Engine was never built, and Lady Lovelace's code was never run.[57] There was little attempt to build a mechanical general computer after Babbage's failure. The task just seemed too daunting.

The field of computing stayed relatively stagnant in the following years. The First World War came and went, with minimal sign that computing was on the horizon. But at the dawn of summer 1912, a boy was born in London who picked up the torch dropped by Lovelace and, at times, even argued with her ghost.[58] By the age of twenty-four, he had revolutionized the field of computer science and restarted Babbage's quest to build a general computer.[59] By the age of thirty-eight, he had published the paper that started the study of artificial intelligence. And at forty-one, he died under tragic circumstances. He was a brilliant flash in the pan of human history – a philosopher, a mathematician, a visionary, and a war hero: Alan Mathison Turing.

CHAPTER 3

War Hero

Alan Turing's life was defined by war. Just two years after his birth, the world was gripped by darkness as the First World War began. Although they were uncommon by modern standards, bombs did fall on the streets of London, where Turing was born, killing numerous children and infants. It's unlikely that he had many memories of the war, as the armistice that ended the bloody conflict had been signed when he was just six years old. But it is a sad truth that the effects of war do not end when the last bullet is fired.

Turing was raised in a Great Britain that was drastically different from the pre-war nation he had been born into. The Great War – as they called it then, unaware there would be a sequel – was different from any armed conflict in history. The soldiers who departed from the British coast had an honorable vision of warfare. What they found was unthinkable, indescribable horror. Horses were replaced by tanks, rifles with machine guns, and tents with rat-infested trenches. The Industrial Revolution provided men with the ability to destroy each other in ways so gruesome that the nations of the world came together in the aftermath in agreement to never use these weapons again. To this day, the idea of chemical warfare remains in the collective consciousness as a crime against humanity so great that it should never be repeated. The men who were lucky to return did not return the same. The Industrial Revolution had changed warfare, and it had taken their sacrifice for those in power to understand what horrors were a step too far.

Although he was raised against the backdrop of the Great War, it had little direct effect on young Turing's life.[1] His older brother was too young to fight, and his father was a civil servant in British-controlled India.[2] The same cannot be said about the Second World War.

Turing's path to the Second World War began during his time at Cambridge. Like Lovelace before him, Turing was something of a child prodigy in mathematics.[3] Not only was he skilled, but he showed a deep love and enthusiasm for the practice. For many boys in Alan's economic position (a member of the British middle class), going to Cambridge meant an opportunity to level up their social standing. Cambridge men were treated with the utmost prestige, no matter what family they came from. Yet for Turing, his social position was of little importance. For him, Cambridge was an opportunity to brush shoulders with the greatest mathematicians of the age.[4]

By all accounts, Turing flourished at Cambridge. Although he had a quirky and sometimes awkward personality, he had many friends.[5] In his classes, he was identified as being gifted, routinely impressing his tutors by solving problems in ingenious new ways.

Yet his undergraduate experience was far from the happiest time of his life.[6] Before leaving for Cambridge, his close childhood friend Christopher had unexpectedly passed away.[7] The death of his friend haunted Alan. His friend's absence sentenced him to a struggle with depression that persisted throughout his undergraduate years. But Alan trudged on, earning an honors degree in mathematics.

Turing's life resembled a powerful river, constantly pushing forward with unstoppable momentum, seemingly unbothered by the obstacles nature put before him. He progressed straight from his undergraduate studies into a master's degree, where his genius in the field of computing began to fully emerge. At the age of twenty-four, he published a paper titled "On Computable Numbers, with an Application to the Entscheidungsproblem."[8] It is widely regarded as one of the most important documents ever written.[9]

Although the "Entscheidungsproblem" was an important puzzle about the limits of algorithms and computation, it is not Turing's answer that is important to our story. What truly mattered was the ingenious method Turing employed to tackle this challenge. Instead of relying solely on mathematical proof, Turing crafted a remarkable thought experiment.

He envisioned and described a machine capable of processing and solving any algorithm that a human could devise. His vision constructed a blueprint for a machine that was mathematically provable to be able to solve any solvable algorithm. Turing's "universal machine," later renamed the "Turing machine," laid the groundwork for the digital computers we use today.

To be clear, the Turing machine was not a physical computer but rather an exceedingly impractical imaginary tool. Its purpose was to help readers visualize the boundaries of computation and algorithms. However, Turing's experiment proved that a machine could be constructed to perform any calculation that a human mind could do – as long as it was provided with the appropriate instructions and had an infinite amount of memory and power. The machine was deemed "universal" due to the idea that it could be programmed to execute any algorithm. One machine: infinite possibilities. A general computer would be the practical application of Turing's design: a machine that could compute any algorithm, given the instructions, within the limits of the memory and power humans could build for it.

What Turing had done was provide form and proof to Lady Lovelace's visions. When he published his work, the world shifted. The scientific ambition to build a general computer that had previously died with Babbage and Lovelace was reignited.

Turing plowed through his PhD at Princeton, producing increasingly inspired work.[10] What was next for this great mind? His head was bursting at the seams with visions of a glorious future. The unstoppable march of his genius had the potential to drag all of humanity forward. Was he to become a professor, working in a highly funded laboratory? Would he teach the next generation of computer scientists at one of England's great universities?

If the narrative of history had taken a different turn, maybe he would have. But it was at this point that the great rapids of war disrupted the flow of Turing's river. The Nazis had risen. The Second World War had begun, and Turing turned his great mind to winning the war.

Alan was a young, able-bodied man. But he didn't fight the Nazis on the front lines with a rifle or in the air piloting a bomber. Instead, he spent the war in a small military communications camp on the outskirts of London: Bletchley Park.[11]

Bletchley was home to a top-secret military operation that Turing had been recruited into. The mission for his team was simultaneously simple and impossible, mundane but vital: Break Nazi codes. Win the war.

During the period, many military messages were transmitted through Morse code on open radio signals, which meant that anyone with an antenna could listen to them. It doesn't take a genius like Alan to know that secret battle plans are not very useful if anyone with an antenna can hear them. To avoid this issue, the messages were sent using a code. German messages were blended and put back together by a device known as the Enigma Machine.[12] The Enigma looked similar to a typewriter, although it had a significant difference. The Enigma had a device that would scramble the letters. If I typed the letter "A," a display system might illuminate the letter "S." One letter at a time, the Nazis would scramble their messages. When the scrambled message was received, it could be typed back into an Enigma Machine to recover the original text. The Enigma operators also had a code book. Each day, the operators would look at the book and see instructions on how to set up the machine to scramble the letters. Every day, the code changed. You needed both the Enigma Machine and the daily code book to unscramble the messages. The Allies had the machine, but they lacked a code book. And even if they got their hands on one, the Nazis could easily distribute a new code book.

Every single day, there were 150,000,000,000,000,000,000 different combinations the Enigma could have been set to.[13] The Nazis were justly confident in their seemingly uncrackable code. It was unthinkable that any human, or any team of humans, could work through the 150,000,000,000,000,000,000 possible combinations every single day. What the Nazis hadn't accounted for was Alan Turing. Turing, as smart as he was, knew that he couldn't decipher the codes himself. But he didn't need to. He was going to build a machine that could: a computing machine. He built the "Bombe."[14]

Like the Analytical Engine, the Bombe was a mechanical device. However, instead of steam, it was powered by electricity. It was designed to execute a codebreaking algorithm written by Turing. The idea was simple. If Turing's team possessed an encoded message, they could guess a set of words that were likely to appear in the message: for example, "good morning" at the beginning of a message. By giving the Bombe the expected result of "good morning" and the actual message received, which might be "xbwi oasrmcv," the Bombe, following Turing's algorithm, could work through the 150,000,000,000,000,000,000 combinations and determine which ones would produce the actual message received. The small number of possible combinations provided could easily be entered by humans into an Enigma Machine until one produced uncoded text.

The system worked marvelously. No matter how many times the Nazis changed their code books, Turing's team was unaffected. They could break any code programmed by the Enigma. Some historians estimate that the invention of the Bombe shortened the war by two to four years.[15] Turing's invention is credited with saving 12 to 24 million lives and was an important part of the Allies' D-Day landing plans.[16] Turing was a war hero, although the British public would not know it. His work was classified, and it wasn't until long after his death that his contribution would be unsealed.

The Bombe, contrary to popular belief, was not the first computer. At least not in the sense that Turing had theorized years earlier. Although technically impressive, the Bombe could only process a single algorithm. It had one goal: to break the Nazi code. It could not be reprogrammed to achieve other tasks. When he was at Bletchley, Turing did work on another machine, the Colossus, which was the world's first digital computer.[17] It was programmable but still narrow in scope, so it is not considered a general computer. Unfortunately for Alan's legacy, the first general computer was designed by the US military in 1945 and used for the much less elegant task of aiding artillery fire planning. After the war, Turing went on to aid in the construction and design of multiple revolutionary digital general computers.

Like most great mathematicians, Turing was a master of patterns. He could intuitively recognize them and use them to solve problems. Are you good with patterns as well? Have you noticed a pattern that this chapter, the story of Alan Turing, has seemingly broken?

At the outset of section one, I promised three stories: an inheritance, the poet's daughter, and a war. Over the last two chapters, when the subject was introduced, there was a large heading for "The Poet's Daughter" and "The Inheritance." But there was none for Alan's triumph during the Second World War.

There is no error. Although Turing's involvement in the world war is of great historical significance, it is not this war that is important to our story. As I said at the beginning, Turing's life was defined by war, and he lived through plenty of them. The war I am interested in was not fought with guns or bombs. There were no enemy codes to be broken. And there was only one casualty. The war I am fascinated by, and the war that would set the path for AI, was Alan Turing's war against death itself.

CHAPTER 4

The War

Christopher Morcom was Alan Turing's first love. They met while at Sherborne School and developed a deep friendship.[1] Turing arrived at Sherborne when he was thirteen years old.[2] His early days at the boarding school were painfully lonely. Boys of that age can be described as cruel in the best of circumstances. They did not care that the young Alan Turing who stood before them was a genius, and had no way of knowing he would become a war hero. All they saw was a weak, socially awkward boy who cared more about science textbooks than rugby.[3]

Alan may not have been able to relate to the other boys his age, but he noticed an older boy who seemed to be just as gifted and interested in the sciences as he was. The other boy, Christopher, was much better adjusted to the social world than Alan and was probably completely unaware of the younger boy's existence.

Turing watched Christopher from a distance for some time, imagining how their friendship would play out if he ever mustered up the nerve to introduce himself. Alan was awkward, and the fact that they were in different years limited the possibilities of a natural meeting. Eventually, curiosity overwhelmed him; he could live no longer without at least trying to befriend the older boy. It is said that Alan approached him with a question about the orbit of planets.[4] There is no doubt that the meticulous Turing would have spent countless hours in his bunk crafting the perfect opening question on something that would perfectly pique Christopher's curiosity.

Christopher's first reaction to Alan's question is unknown. But in the long run, Alan's plan was a success. One of their teachers reported that, during a half-term game of football, Christopher restarted the discussion on the orbit of planets.[5] It quickly became clear that the two were meant for each other. They shared uniquely inquisitive minds and pushed each other to pursue increasingly more knowledge.[6] Much of their free time was spent together in the science labs, designing experiments and exploring the world of science.

It was clear to everyone at Sherborne, both teachers and students, that the two shared a special bond. What was impossible for them to have known was that Alan's feelings grew beyond friendship. Over their time together, Alan fell deeply in love with Christopher. It was not the juvenile boyhood love usually associated with kids their age. In Christopher, Alan had found a soul that matched with his. Christopher inspired him and made him feel safe. But a deeper exploration of their relationship was not written in the grand script of the universe.

During their friendship, Alan had often noticed periodic unexplained changes in Christopher's health. He never enquired why, and Christopher never addressed it. Unknown to Alan, Christopher had been keeping a deadly secret. As a child he had contracted tuberculosis, and his fight with the disease had left his body irreparably tattered.[7] Every day was a blessing for him, and he knew his days were probably numbered. He chose to spend those days with his friend Alan, asking questions about the nature of the universe.

Alan was caught completely off guard by Christopher's death. When he was told, he wrote a letter to Christopher's mother: "I am sure I could not have found anywhere another companion so brilliant and yet so charming and unconceited,"[8] he wrote. "I should be extremely grateful if you could find me sometime a little snapshot of Chris, to remind me of his example and of his efforts to make me careful and neat. I shall miss his face so, and the way he used to smile at me sideways."[9]

Alan was seventeen years old. He began his studies at Cambridge a year and a half after Christopher passed.[10] There have been numerous academic papers written on the effect of losing a friend at this point in one's life – the effects of having to face the truths of mortality so intimately. But no study can capture what Turing experienced during that time, after having the naivety of youth shattered. Only those who have lived through such an experience can truly comprehend. Unfortunately, Turing and I share a tragic connection. The summer before I left home for my bachelor's degree, I received a phone call that would change the trajectory of my life. A dear friend and personal hero of mine had passed away. Like Christopher, he had hidden his struggle with an invisible enemy and suffered in silence.

Like Turing, I was shocked by the news and debilitated by an initial period of shock. But while shock and the sympathy of others quickly dissipate, depression lingers. Suddenly you find yourself in an alien environment and are left to your own devices to make sense of what you have experienced. No matter how hard you try to pretend that nothing is different, everything has changed. The world becomes a different place, and you need to make sense of it if you are to carry on.

Alan struggled in much the same way I did. Reading his writings, I can feel his pain. He had lost the one person who understood him, his only true friend, and he felt alone. But more than that, he felt out of touch with the world around him and out of control. Losing a contemporary at that age is different from losing a grandparent or even an adult aunt or uncle. Although both are tragedies, there is something uniquely painful about the loss of a youthful soul. It shatters our illusions of how the world ought to be. It feels like a crime, a stolen unrealized life, a soul deprived of the joys of living. It produces the type of pain that keeps one up at night and, if left unchecked, can corrupt one's soul.

While at Cambridge, Turing searched for answers. Between his classes in maths and science, he pondered questions of a more spiritual nature. He longed for answers to unanswerable questions. He was fighting an unwinnable war in his own mind and in his heart. An atheist, Alan searched for his own answers about the fate of his friend after death. It was in this battle, a battle to understand the nature of life and death itself, that Turing laid the foundations for his revolutionary philosophy of artificial intelligence – the philosophical foundations of the seventy-three-year journey that has created our current moment, the journey that has made this book necessary.

In his pain, he asked questions about the existence of the human spirit. Two years after Christopher's death, Turing visited Christopher's parents, with whom he had maintained a close relationship.[11] In his notes, he speaks fondly of his time with the Morcoms. But he could not ignore the specter of Christopher that lingered in his parents' home. Something in the house possessed Turing to write a short essay called "Nature of Spirit." He wrote:

As McTaggart shows matter is meaningless in the absence of spirit (throughout I do not mean by matter that which can be a solid a liquid or a gas so much as that which is dealt with by physics e.g. light & gravitation as well, i.e. that which forms the universe). Personally, I think that spirit is really eternally connected with matter but certainly not always by the same kind of body. I did believe it possible for a spirit at death to go to a universe entirely separate from our own, but I now consider that matter & spirit are so connected that this would be a contradiction in terms. It is possible however but unlikely that such universes may exist.

Then as regards the actual connection between spirit & body I consider that the body by reason of being a living body can "attract" & hold on to a "spirit" whilst the body is alive & awake the two are firmly connected & when the body is asleep I cannot guess what happens but when the body dies the "mechanism" of the body, holding the spirit, is gone & the spirit finds a new body sooner or later perhaps immediately.

As regards the question of why we have bodies at all; why we do not or cannot live free as spirits & communicate as such, we probably could do so but there would be nothing whatsoever to do. The body provides something for the spirit to look after & use.[12]

His words give us a glimpse into his thoughts on the nature of existence. During this time, he developed a belief in a spirit that existed separate from the body, which meant that, even though Christopher's body had died, his spirit in some way remained.[13] In letters to Christopher's mother, he would often speak of the peace the thought of Christopher's spirit gave him. But in "Nature of Spirit," Turing subtly expresses an idea that would define his later work in computing. When pondering the purpose of the body, Turing states that the spirit could live without a body but doing so would be useless as there would be nothing to do. For Turing, the spirit holds the soul, the essence of the person, and the body is a machine through which the spirit can operate. The ability to think was not a function of the spirit itself but of the physical brain inside our skulls. The spirit simply acted through the physical machine of the brain. He would dedicate his later life to building similar machines.

As Turing grew older, he had a front seat to the development of early digital computers. The systems he worked on were weak by today's standards. But visionaries are not limited to thoughts on what is sitting in front of them. Like Lovelace examining the plans for the Analytical Engine and foreseeing the advances of Turing's time, Turing recognized the exponential potential of computers. Instead of just wondering how powerful they could get, Turing turned his mind to comprehending a more philosophical question: Could machines think?

His answer to the question was an unapologetic "Yes." In 1950, he published what would become the founding document of artificial intelligence.[14] I have always found it charming that, although Turing was a Cambridge man, the paper that became his most influential was published by Oxford, Cambridge's bitter rival. Oxford's journal *MIND* is one of the most prestigious philosophy journals in the world. It was undoubtedly an odd place for the world's leading computer scientist to submit his work. One can only imagine the puzzled look on the editor's face when a manuscript from Alan Turing, the "father of computing," arrived on his desk. "What does this mathematician think he knows of philosophy?" he may have cried aloud to his colleagues, chuckling to himself before opening the manuscript and beginning to read. How fast might his expression have transitioned to one of wonder?

Turing approached the question in a way no one before him had. He begins with two simple questions of his own: What do we mean by a "machine"? What do we mean by "to think"? The first of these two questions was relatively easy for him to answer. The machines he would be referring to were digital computers, although, as he notes, a mechanical computer like the one designed by Babbage could theoretically perform the functions he was discussing (although it was not at all practical).

The idea of "thinking" presented a more difficult challenge for Turing. How does one define the ability to think? Turing turned the question on its head, suggesting that the concept of thinking was not one for which we could create an objective standard. He believed there was no objective way to define what it meant to think. Instead, he would design a game that only an intelligent being could win. He called his invention "The Imitation Game."[15]

The game, at its most basic level, is quite simple. There are three players: the interrogator, the human, and the machine.

The game is text-based and has no set time limit. The interrogator, who is also a human, questions the two subjects with the goal of figuring out which one is a human and which one is a machine. The human's job is to prove that they are human. The machine wants to convince the interrogator that it is the human.

The human wins the game if the interrogator can correctly identify them.

The machine wins if it fools the interrogator into believing that it is the human.

But winning once is not enough for a machine to prove that it is a thinking machine. The machine needed to prove, over multiple games, that it could render the human interrogator useless at detecting who the machine is. This idea became known as the Turing Test.[16] Turing thought that, in fifty years' time (the year 2000), digital computers would be powerful enough to pass his test.[17] He then continued his paper by explaining how digital computers worked and presenting the evidence for why he was certain that a machine would soon be able to pass his test.

The paper made it clear to all who read it that the greatest computer scientist of the age was convinced that machine intelligence was on the horizon.[18] But acceptance of Turing's hypothesis relied on one vital assumption. He had changed the question from "Can a machine think?" to "Can a machine pass a Turing Test?" There was little debate over whether or not a machine would be able to pass a Turing Test. The real question was, did the Turing Test really test intelligence?[19] Turing's test, and his belief in the possibility of intelligent machines, has received much criticism over the years. It is my belief that much of the criticism is born from a narrow understanding of Turing's philosophy. It is a philosophy that only fully reveals itself when the effects of Christopher's death are taken into account.

In his paper, Turing admits that he has no positive proof that the test can measure intelligence or the ability to think.[20] He didn't necessarily see that as a problem, since he believed the idea of intelligence itself to be arbitrary. It is naturally unmeasurable. Turing never meant to say that his test could measure all forms of intelligence. Instead, he believed that the concept of intelligence was an emotional one produced by humans.* Something is intelligent because it behaves in a way that others think is intelligent. By this measure, the Turing Test is not actually examining the machine's ability to fool a human but rather the human's ability to believe that the machine is intelligent in the context. The example of the Imitation Game was a specific human context. The machine is intelligent because of its ability to adapt to human questions and provide intelligent answers.

At some point while writing this argument, a ghost approached Alan. Our old friend Lady Lovelace spoke from beyond the grave. She presented a fierce rebuttal to the theory. If you remember, Lady Lovelace thought similarly to Alan, but she stopped short of him, stating outright that a machine could never be creative: "*The Analytical Engine has no pretensions to originate anything. It can do whatever we know how to order it to perform.*"[21]

Turing referred to this statement as "Lady Lovelace's Objection," and he refuted it with terrifying precision.[22] For Turing, the ability to learn was of the utmost importance to his idea of

---

* I have relied heavily on the research of Diane Proudfoot to make these points.

intelligence. He proposed that the Analytical Engine that Lovelace observed, unlike his digital computers, had been too rudimentary for her to observe the possibility of a machine that could learn. He presented the idea that a digital computer could be programmed with an algorithm that gave it a reflexive mechanism, similar to the one biological creatures have, which would allow it to learn. This algorithm would allow it to combine information and past experiences in order to create new outcomes.

If such an algorithm were to exist, a machine could learn from the information given to it by humans and create outputs that were uniquely its own. If such a theoretical algorithm could be written, then why could a machine not create something new or surprising? Turing often faced criticism from colleagues who believed that the computer he was working on was nothing like a human brain, as it could only answer problems given to it.[23] They also argued that it could only answer those questions based on instructions given by the human operator. For example, a chess algorithm run by Turing's computer could only play within the bounds of the moves already imagined and played by human players. Yet Turing firmly believed that, one day, his machines would be able to learn and that the secret to machine intelligence was not to try and build an adult robot but to build a childlike one that would make mistakes and learn from them. If he succeeded, he could create a chess algorithm that would not simply move the pieces as a human would but could learn the rules of chess and create its own moves.

Turing died before he could build a machine that created its own chess moves. But in 2017 an AI system made by Google, trained through the method of computer learning that Turing had imagined half a century earlier in his rebuttal to Lovelace, revolutionized the game of chess. Given nothing more than the rules of the game, it created new strategies that had never before been conceived of by a human.[24] The machine is now unbeatable, even for the world's greatest players. It is too creative and too cunning. Turing seemingly saw the future once again.

To understand why Turing could see this future so clearly, we must revisit his theory of life itself. Turing is often accused of suggesting that he could create a machine that was the same as a human. This idea would imply that humans are simply machines. We are only following our programs after all. Those who oppose the idea that we are just complex machines, no different from a computer with no free will, viciously reject Turing's thoughts. Furthermore, they use the argument that humans are not machines to discredit the idea that machines can be intelligent. Those on the opposite side of the debate believe that, one day, we will be able to create machines that are identical to humans, as the essence of human existence is in our physical existence. We are made of matter and programmed by our DNA. Why could we not recreate this pattern with a machine? A body of silicon and a mind of code. But both of these arguments fundamentally misunderstand Alan Turing.

As he states in "Nature of Spirit," Turing believes that the body is a machine. But he also speaks of the importance of the spirit that inhabits the body. Yes, the ability to think may be given to us by material matter, whether it be the squishy pink goo of our brain or the silicon of a computer chip. But whatever gives us the ability to love, to be ourselves, or to wonder, that exists outside of the body. So there must be an external driver for the vehicle of man: the mysterious soul.

For Turing, a machine that passes the Turing Test is not a human. It is not even like a human. If it were asked to write a poem about love, it could write it.[25] But not in the same way a human would. It would not be drawing on the lingering betrayal of a lover's broken promise to inspire its words. It would be mimicking what a human would write. It can understand what is being asked of it, but its product is soulless, whether or not we can detect it. The machine is imitating the poems it has read – *however, that does not mean it is not intelligent*. A young student may write their poetry assignment, mimicking Lord Byron, despite never having fallen madly in love themselves. They may use the words "passion," "eternal," or "rapturous" without truly understanding the experience that the words denote. But is this student not intelligent? Have they not performed an intelligent act – even if some may not call it "art"?

Turing could answer the question of machine intelligence because he had already separated the soul from the machine. In doing so, he was left with no mystical delusions, only the physical limitations of his inventions' implementations. As we explore artificial intelligence together, we must remember to follow his lead. Turing sought to explore the limits of how far we could push computing by examining ourselves. He was asking what makes our biology so effective and how to replicate it in machine form. He never intended to build a human machine.

The Imitation Game invited the world into Turing's mind. He knew what these machines would be capable of. The Imitation Game was a theoretical warning of what would be possible and a call to action for computer scientists to realize the potential of their work. Philosophical debates over whether the machine is truly "thinking" are of little importance when the machine is successfully completing its objective. It need not be a human machine but rather a machine that can learn so much about humans that, under the right conditions, it can fool us. We should not think of AI as a human-like being but as an algorithm, created by humans, that can act in intelligent ways. Some systems may mimic us at times if we ask them to. But we are different. AI is simultaneously the human-like avatars created by ChatGPT and the abstract algorithms that silently trade on the New York Stock Exchange every single day. It can feel at once human and at the same time extremely alien.

So what is AI?

I have fallen in love with a definition of intelligence that stems from Turing's work – a definition that completely removes the human from the equation. Stuart Russell is considered to be a pioneer of AI. His textbooks are the educational foundations for many of the programmers building AI systems today. He defines AI as a synthetically created system (such as a computer) that acts in an intelligent way to achieve its goals.[26]

Unlike Turing, Russell provides a more specific definition for "intelligence." According to him, something is intelligent when it can perceive its environment and, based on its perceptions, change its behavior to better achieve its goal.[27]

When I discuss AI systems, this definition is what I am referring to. I prefer this definition as it decenters humans from the equation. Intelligence should not be determined based on a human standard. If we were to judge a bumble bee's intelligence by its score on the SATs, it would appear to have no intelligence. However, I am not sure I would fare much better at trying to figure out the optimum way to collect pollen for honey production.

An intelligent system can be simple, such as a microorganism in our gut that spreads its arms when it detects sugar,[†] or it can be extremely complex, such as the management decisions made every day by humans.

As we will explore, AI is nothing new. In fact, my mother worked on AI programs in the 1980s. We have long been able to make machines that act in "intelligent ways." But their functions have up until now been very narrow.

AI hasn't come out of nowhere. It has been developing since Turing published his paper and sparked the ambition. But there is a difference now. In 2012, unbeknownst to most, the world

---

† An example used repeatedly by Russell.

changed, and Turing's vision became a reality.[28] After a decades-long battle, a team of researchers finally demonstrated the practicality of a machine learning technique called deep learning. It is effectively the reflexive mechanism discussed by Turing, the algorithm that gives computers the ability to learn independently. In a flash, AI went from being very simple to potentially extremely complex. We are now dealing with the consequences.

ACT 2

The Chess Game

How We Got Here

CHAPTER 5

Setting the Board

"Work hard, and if you do ... maybe someday you will be almost as successful as I am," Howard would say to his son Geoffrey as he dropped him off at school.[1] Geoffrey was never allowed to forget the success of his ancestors. His father was a well-respected insect biologist, better with bugs than with people. His great-grandfather had been a famous mathematician and science fiction writer. His great-uncle had invented the Jungle Gym.[2] His second cousin, a nuclear physicist, had been one of the few women to work on the Manhattan Project. And his great-great-grandfather had been George Boole, one of the most important mathematicians to ever live. Needless to say, expectations were high.

To Geoffrey's credit, he wouldn't become almost as successful as his father – he eclipsed him. In academic conversations, Geoffrey Hinton's name is often spoken with the same respect as his great-great-grandfather's. His story is so important that he is often introduced as the godfather of AI. But Geoffrey's story is not a smooth one. He spent most of his career on the outside looking in. Some of the same people who now celebrate him once dragged his name through the mud and painted him as a failure.

I like to imagine the story of our march from Turing's paper toward the current moment through the metaphor of a chess game. Imagine you are sitting at a chessboard. The pieces are set, and the rules are written. This set-up was the condition when Geoffrey started his journey into AI research.

Turing, Lovelace, al-Khwārizmī, and countless others had set the stage for AI. The pieces had been crafted and put into place. Now, they just had to be skillfully moved so that one player could take the king. Some pieces in the game were more important than others. But they all needed to work together to achieve a common goal. One may think of Geoffrey's role as that of the queen. She is a powerful piece, often active from the beginning, although not always present in the early battles.

There are three phases to any chess game: the opening, the middle game, and the end game.

In the opening, the first pieces are moved and then developed. The player primes the board for the game they are preparing to play. The opening is the shortest phase as players execute their opening strategies with minimal opposition.

In the middle game, the players continue to develop their major pieces, while simultaneously setting up their attack and testing the enemy's defenses. In the middle game, you may try several strategies, only to see them fail and have your pieces taken. The middle game takes patience and the ability to plan for the long term. This phase is where most chess games are won and lost.

If both players survive the middle game, they progress to the final stage. The end game occurs when the board is mostly cleared of pieces. Vicious attacks have been mounted to no avail. Your defensive line has held but at great cost. If you played your moves right, you should begin the end game with better pieces than your opponent. Now is the time to mercilessly hunt down their king and win the game.

Although each section of the game is unique, the lines between them are often blurry. A player often knows when they are in the middle game or the end game, but probably couldn't tell you the exact moment when they started. The game of chess, like life, is blurry and complex. The road to contemporary AI is littered with fallen pawns and broken bishops. The game was not a straightforward one. Turing's predictions about the speed of AI development were wrong. We made countless mistakes.

A careful reader may note that there are numerous alternate universes where our decisions led to defeat – where we might not have been able to develop the powerful AI we have today.

CHAPTER 6

The Opening

It didn't take long for scientists around the world to take up Turing's challenge. To be fair, some moves had already been made before Turing's work had been published. The world of computing was abuzz with excitement, and it offered fertile pastures for academic curiosity. There were no set rules, and accepted standards for the best way to build "thinking machines" had not yet been established.

There is a common myth about knowledge: that it continually builds and increases. Someone has a new idea; we use science or logic to prove its validity; and it is then added to our body of knowledge. If we are careful and follow the rules of knowledge-building, over time we will increasingly improve our knowledge of the world, forever progressing toward an objective truth. This was the romantic idea of knowledge popularized during the Enlightenment, the period when we developed our modern academic institutions. However, it doesn't take long to see that this idea is a myth once you are inside the system.

This is not a novel observation. There are massive bodies of research dedicated to understanding the failings of the Enlightenment's promises. In support of these theories, there are numerous studies that explore the realities of knowledge-building. Even the historical stories about the discovery of knowledge that we now accept as true are often quite a bit messier than the way they are presented. In my favorite book on the topic, *The Structure of Scientific Revolutions* by Thomas Kuhn, major advances in science are completely reframed.[1] When Einstein proposed the theory of relativity, it did not simply build on the previous science. Instead, he completely rejected the accepted rules and laws of physics. The Newtonian model that had dominated physics for two centuries was completely wrong according to Einstein. It couldn't provide answers for the new questions that physicists were asking.

What Einstein did was build a new "paradigm," a new set of rules and assumptions that built the base from which other researchers could work. The transition to Einstein's new paradigm was not smooth. Even as evidence mounted that his assumptions were better suited to providing answers to the observed natural patterns of the universe, the old paradigm persisted. Thousands of researchers would die having never accepted Einstein's accomplishments.[*] And once his theories were accepted as the dominant paradigm in the field, much that had been accepted as knowledge by the previous paradigm was discarded as misguided and useless.

But even Einstein's work was not the end. His theories laid the foundation for better research, but eventually there were new questions that could not be answered within his paradigm, and new ones emerged.

Academia is not a harmonious place. It's a dogfight. There are numerous paradigms fighting against each other at any given time in every field of study. There are dominant paradigms that we work within to help us develop new knowledge, but most of

---

[*] It must be said that Einstein's theory faced far more criticism than it deserved. Although there was much valid criticism from prominent scientists, there was also a large movement against his work driven by antisemitism in the scientific community.

us are well aware that a lot of what we believe is probably wrong or incomplete. This is not a problem – it is the true function of knowledge creation. The job of the academy is to interrogate our beliefs, much like a Turing Test, and try to decipher which are true and which are not.

Working within an accepted or dominant paradigm has some major advantages. For one, there are a lot more resources available. Universities and governments have very limited budgets. And there is no shortage of good ideas on a university campus. While disruptive science is also funded, as it is recognized as being an important part of the discovery process, the lion's share of money goes to those working within a paradigm that has been proven to get results.

I referred to the opening period for the game of AI as "fertile pastures," since the academic discipline was new. There were no kingpins who had erected fences. Walking paths had not yet been built, and the soil had not been tilled. There were endless possibilities, and the early researchers enjoyed a freedom that allowed them to explore any direction they desired. Some directions were complete dead ends. Others would turn into the dominant competing paradigms for the development of AI. Unfortunately, I cannot explore all the twists and turns that this journey took, and my narrative will be somewhat simplified.

The first major move occurred only a year after Turing published his paper in *MIND*. A computer scientist by the name of Marvin Minsky built a machine designed to mimic the human brain.[2] His device, known as SNARC,[3] was not designed to mimic the entire brain but instead just forty neurons.[4] Neurons are the cells that form the basic building blocks of the brain, responsible for receiving sensory input from the external world and sending motor commands to our muscles, as well as thought and planning. Minsky was inspired by the work of a neuroscientist and a logistician named Warren McCulloch and Walter Pitts, respectively. In 1943, they published a theory of how the function of a biological neuron could be mathematically replicated.[5]

The human brain is made up of about 86 billion neurons. Even more exist throughout your body, connecting your senses to your mind. Their individual function is relatively simple. A single neuron will receive a message in the form of an electric signal from the other neurons it is connected to. If the neuron receives a strong enough message, it will pass the signal on to all of the neurons it is connected to. And the chain continues. Each neuron has a binary choice: pass a message on or don't pass it on. The strength of the signal must reach a biological threshold in order for the neuron to pass on the message. We learn when our body strengthens connections between specific neurons, making messages between them stronger and creating pathways that define our actions.

For example: Imagine you are in a room full of boxes. Inside half of the boxes are fart bombs. Inside the other half are candy. You have no clue which box contains which. However, you have *nine* friends with you who saw the boxes get loaded. Some of your friends want you to get the candy. Some of your friends just like to watch the world burn and want to see you open the fart box. And some of your friends weren't really paying attention when the boxes were loaded and aren't 100 percent sure which box has what. Your friends can't talk to you; all they can do is raise their hands to vote on whether you should open each box.

At first, it would make sense to go with the majority. If *five* of your friends raise their hands, you open the box. But pretty soon you realize that you seem to be getting fart bombed more than you like. And you notice that one of your friends, let's call him Reggie, almost always seems to vote yes on the boxes with candy. So, you start giving his vote more weight. Suddenly, a box that only has *four* votes, but one of them is Reggie's, is enough for you to open the box. Unlike Reggie, your friend Dock appears to be pretty useless. He just wants to watch the world burn and is telling you to open random boxes. You decide not to listen to him at all. Finally, you notice that your friend Sheen seems to be pretty trustworthy as well. He seems to be trying to help you out but definitely wasn't fully paying attention when the boxes were loaded. You decide to weigh Sheen's vote higher but not by quite as much as Reggie's.

By the end of the game, you can confidently choose which boxes to open and which ones to ignore based on how your friends vote. It is not a democratic decision but instead a vote based on different voting weights, assigned to different friends. This process is exactly how our neurons work when we are learning. When outcomes are good, a neuron builds stronger bonds with the neurons that sent it the message.

McCulloch and Pitts were inspired by Turing's early work, where he first demonstrated the potential power of computing machines.[6] Building from his theoretical Turing machine, they demonstrated how mathematical (Boolean) logic could theoretically be used to create an artificial neuron that could be run by Turing's hypothetical computer.[7] Boolean logic is a way to make logic-driven decisions using only "yes" and "no" or "true" and "false" propositions.[8] It has become the backbone of every contemporary coding language, as it allows us to model complex algorithms in the binary code used by digital computers. If you recognize the name "Boolean," that's because it was invented by George Boole, Geoffrey Hinton's great-great-grandfather.[9]

Marvin Minsky's machine didn't use Boolean logic and was quite rudimentary. It was designed to take the place of a rat in a maze, using vacuum tube circuits to crudely recreate the function of a series of neurons.[10] Each vacuum tube took the place of an individual neuron, randomly connected to other vacuum tube neurons. Signals sent through the interconnected vacuum tubes were used to decide which route to take through the maze. When the machine correctly navigated the maze, control knobs were activated that increased the probability that an electric signal would be sent down that path again.[11] Although it did not perfectly simulate a neuron, the machine had some ability to learn from past experiences in the maze. Although it was pretty dumb, it was acting with intelligence.

In 1956, Allen Newell, Herbert Simon, and Cliff Shaw took a different approach to creating AI.[12] Instead of taking inspiration from human biology, they took inspiration from philosophical theories on how humans think. They were uninterested in the physical structure that created the end product of thinking but instead focused on the result itself. The approach taken by these three researchers would rely on a logical system. You would provide the program with an input, and it would run the input through a predesigned computer algorithm. This algorithm would interpret the input based on a set of precoded logical rules. This system relies on humans developing rules about the relationships between subjects and teaching those rules to a machine.[13] The theory relied on the idea that we could, with enough brute force, model the world as a series of logical statements that could be understood by machines. To explore this idea, let's return to the fart box game.

This time, whenever my friends vote on a box, instead of weighting their votes based on the outcome, I consult a rule book given to me at the beginning of the game. After a 5–4 vote, I open my first box with a fart bomb in it. I look at the rules and consult them for my next move. The rules, which refer to my friends as [players] {1–9}, say that, logically, the players who voted "yes" voted on a box with a fart bomb, therefore I should only open a box that receives votes from the four players who didn't raise their hands. I continue this process, opening several boxes without problem. Even when I open a good box, I make sure to consult my rules.

Based on which of my friends voted on each box, the rules for how I should respond to each box change. In this example, I will still change my behavior, resulting in me choosing better boxes. However, I am not personally learning. I am executing logical rules or instructions that are prewritten and given to me. In this system, the actions and objects of the real world are replicated on paper through symbols that you interpret in order to act in an intelligent manner. The rules disregard the names or personalities of my friends and present them only as symbols on paper that can be logically manipulated: [players] {1–9}.

Newell, Simon, and Shaw used this technique to develop a program called the Logic Theorist.[14] The Logic Theorist was designed to provide "proofs" for mathematical problems. A proof is a list of logical statements that demonstrate the validity of a mathematical statement. The program worked brilliantly, demonstrating its ability to create mathematical proofs at a level on par with some of the world's top mathematicians.[15]

The success of these innovations led to the assembly of the first AI conference. The eight-week conference, held in Boston in 1956, was to be a giant brainstorming session that would bring together the greatest minds of early AI research.[16] At the time, the field had minimal direction, as it was still very young. John McCarthy, the event's main organizer, dreamed that the conference would unify the researchers around a single research direction into which they could invest their energy. He wanted to solidify a paradigm.

Instead of unifying the researchers, the conference became a mess of ideas.[17] The researchers met every day and discussed their theories and findings. Grand conversations were held exploring the potentials of this new field of computing. New methods were shared and debated, but no consensus was reached.

What the conference did do was solidify artificial intelligence as a field of study. Up until the conference, there had been no community and no set goal. Instead, there were just a bunch of professors, from different academic backgrounds, scattered around the world. In the 1950s, there was no email, and long-distance phone calls were expensive and inconvenient. The conference brought together a group who could begin to form the knowledge base for the study of AI. It was these men who would get to decide what AI was and which research paths were the best ways to achieve it. In fact, before this conference, the term "artificial intelligence" did not exist. It was at the conference that the name was invented, and a new field of research was born.[18]

Although the conference did not create a single direction as McCarthy had hoped, multiple paradigms did begin to emerge.[19] The two most important paradigms are represented by our two box games. In one paradigm, the researchers create an environment and structure for the machine to teach itself. This is the **neural network** method. In the other, the researchers construct logical code that a machine can execute. For the machine to act intelligently, the world must be interpreted as symbols that the machine can read and run through its logic-driven system. This approach became known as **symbolic artificial intelligence**.[20] The history of AI development, from this point forward, can largely be told as a conflict between these two paradigms – a battle for recognition, funding, and academic legitimacy.

CHAPTER 7

Let the Middle Game Begin

With the field of AI established, numerous exciting discoveries appeared to be just over the horizon. One year after the conference, American psychologist Frank Rosenblatt presented an astounding improvement on Minsky's SNARC system.[1] Unlike Minsky, who had designed a machine that simulated several neurons, Rosenblatt decided to increase the scale of his machine but narrow his objective. Rosenblatt's machine was to be far more high tech, but instead of trying to make forty interconnected neurons, he wanted to make one.[2] One single, artificial neuron.

A psychologist, Rosenblatt was much more interested in understanding how the brain worked than in creating a "logical machine." He had read McCulloch and Pitts's work and become fascinated with the concept of simulating the function of the human brain. He thought that successfully building an artificial neuron might aid us in better understanding our own brain.[3]

Rosenblatt's 1957 machine, the "Perceptron," was a colossal beast.[4] It worked in a very similar method to our neural network box game. Instead of friends voting, the Perceptron was hooked up to a camera with 400 photocells.[5] Each photocell would be the equivalent of a pixel for today's digital camera. The idea was that the neuron would play a game, similar to the box game, where it needed to make a binary choice. For example, Rosenblatt would show the camera an image of a triangle or a circle. To win the game, the Perceptron needed to correctly identify if it was looking at a circle or a triangle. It would make its choice by turning on a light if it saw a triangle or leaving the light off if it saw a circle.

Imagine that each photocell (or pixel) hooked up to the Perceptron were a friend voting on whether or not what they were looking at was a triangle. Based on what appeared in each photocell, such as more or less shading, the photocell would send an electric signal to the Perceptron. If a strong enough electric signal was sent, a large light would turn on, proclaiming "I see a triangle!" At first, these signals meant nothing, and the Perceptron was basically randomly guessing whether or not to turn its light on. However, when it guessed right, Rosenblatt would press a button telling the machine "Good job! That was a triangle!" When the machine received positive reinforcement, it would strengthen the weights for the pathways that related to the good outcome. It would start weighting the electric signals of some photocells more than others.[6] Eventually, the Perceptron proved itself to be excellent at identifying triangles and circles. It was a massively important proof of concept for the supporters of neural networks.

Following his triumph, Rosenblatt declared:

Stories about the creation of machines having human qualities have long been a fascinating province in the realm of science fiction. Yet we are about to witness the birth of such a machine – a machine capable of perceiving, recognizing and identifying its surroundings without any human training or control.[7]

The expensive Perceptron project had been funded by the US Navy, who was equally excited by the machine's success.[8] Unfortunately, the celebration would be short lived. The Perceptron proved to be extremely limited in its abilities to recognize patterns. Although it performed well with simple shapes, the complex patterns of the real world were far beyond its abilities. This issue was not just a problem of hardware but also one born out of its theoretical structure. McCulloch and Pitts's system was just too simple.

The loudest critic of Rosenblatt's work was an unexpected one. Marvin Minsky, the man who built the first neural network, publicly and fiercely tore down Rosenblatt's accomplishments.[9] Minsky, drawing from his experience with the SNARC system, declared that Rosenblatt was far too bullish on neural networks and that his passion for the architecture of the brain was blinding him to the technique's limitations.

Minsky's criticism was a deadly blow to the neural network approach.[10] Although the method showed early promise, it failed to live up to the hype. The Perceptron project faltered and was eventually defunded. The harsh criticism from Minsky, not only a respected scientist but also a pioneer of the approach, mixed with experimental failures, made research proposals using the neuron approach toxic.[11] The first attack of the chess game had failed, and the paradigm of symbolic AI now took center stage.

Fortunately, the concept of neural networks did not completely die with Minsky's attacks. There were those who wholeheartedly believed that symbolic AI was a dead end.

"If we are trying to build an intelligent machine, we must construct it the way intelligence is built in nature," the neural researchers would proclaim. "The brain functions using a neuron system, so an intelligent computer should as well."

On face value, this argument made sense, but it was quickly rebutted by the supporters of symbolic AI.

"When we were trying to build flying machines, we wasted years engineering planes with flapping wings, only to watch them crash time and time again," they responded. "We must accept that human technology and the hand of god are not one and the same."[12]

And just like that, symbolic AI took center stage.[13] It had its moments and found some success over the following fifty years. But it would never live up to its expectations. There were some cases where the symbolic approach was practical, but it struggled when scenarios were blurry.[14] The reality is that the world is just too complex to be logically modeled by humans. We are constantly dealing with problems that require massive amounts of background information to comprehend them, let alone answer. Even if a logical system could be made to represent the world, it could only capture so much, as the social world is constantly changing. In fact, the social experience of the world can be drastically different just between two cities, let alone nations. Symbolic systems thrived in small, defined settings but were far from the AI that Turing had dreamed up. They could often be intelligent but not very useful.

Over the coming decades, AI would go through a series of winters and summers.[15] Progress would stall, and funding would dry up; then a new breakthrough would catch the public imagination, and more money would appear.[16] IBM trained a computer, using symbolic AI, that beat Gary Fisher at chess, and the world was captivated.[17] When would such powerful technology be available to them? Robot butlers *must* be just around the corner. But the hype never seemed to materialize into true progress.

Looking back, the greatest innovation of this time may be something that most programmers wouldn't even refer to as AI but that went hand in hand with it. The concept of conventional machine learning is heavily associated with AI but, depending on one's definition, is not necessarily AI.[18] Machine learning (or ML for short) is in many technical ways a midway point between the symbolic and neural approaches.[19]

Like the neural approach, an ML system learns about the world through trial and error. Like neural networks, it runs data through its systems in order to identify patterns and provide answers. However, unlike a neural network, the structure of the system is designed by humans. In short, a human tells the system what it should be looking for. This process is similar to the symbolic approach, as the way that the machine should interpret the world is defined by the programmers.

For example, if scientists are using traditional machine learning to identify different kinds of trees, they tell the system what it needs to look for, like the shape of leaves or the color of bark. The program will then examine thousands of pictures of trees, paying particular attention to the bark color and leaf shape; through a trial and error process, it will learn to identify trees based on their bark color and leaf shape.

In a neural network, the system learns to identify trees on its own, figuring out which features of the trees are key. It only needs photos of the trees. We do not need to tell it to look at variables such as bark color or leaf shape. Through altering the node connections, it will identify these variables on its own.[20]

I suggest you read over the last two paragraphs again. Maybe even highlight them. As mundane as it may seem, these paragraphs will become the root of many of our problems.

When we discuss algorithms, we are typically talking about ML algorithms. They have been prominent parts of our lives for many years, starting before the current AI boom. For example, Google Maps uses such algorithms. Most video games and social media do as well.

Although these algorithms could "learn," they are not intelligent in a way that should raise alarm. They need information force-fed to them. They can recognize patterns but only within the narrow structures given to them by humans. However, they did help to lay the groundwork for Geoffrey Hinton's explosive and triumphant return.

Like I said, the middle game is about patience. It is about not chasing fads and laying foundations for a later victory. And Hinton was nothing if not patient.

CHAPTER 8

The Hippie at My Door

In the fall of my sophomore year, I was rudely awoken by the philosophy-loving hippie who lived down the hall. He kicked down the door to my dorm room and violently shook me awake.

"David!" he shouted. "I just read the greatest book ever, and I need to tell you about it!"

I sat half asleep as he rambled for an hour about *Tractatus Logico-Philosophicus*, the seminal work of Austrian philosopher Ludwig Wittgenstein. I would love to tell you what the book is about, but honestly I have no clue. It was 4 a.m., and I wasn't really listening. "This is just a phase," I thought. "Next week he will be obsessed with some other obscure philosopher." But I was wrong. That philosophy-loving hippie, one of my closest friends to this day, never got over Wittgenstein and has dedicated his life to understanding this obscure philosophical thinker.

This story is not about Wittgenstein's contributions to AI. Honestly, I don't understand his work, no matter how many times I have unwillingly been held hostage by lectures on the topic. Instead, I am intrigued about the decision to study something with no obvious value. This comment is not me judging my friend. I deeply respect the work he does. But he is the first to admit that, as of now, his work is beyond abstract and is referred to by many as "philosophy for philosophy's sake."

One night, as we discussed the pursuit of "useless knowledge," he presented me with a thought experiment. He asked me to imagine building a time machine and traveling back to nineteenth-century Europe. "Ask every professor what the most useless research at the time is – what the most useless research in the world is," he said. "Every professor would have the same answer!"

It is likely, my friend went on to say, that they would have said that the brilliant George Boole was wasting his time and university resources. Boole was developing a strange new form of notational logic – a form of mathematical philosophy.[1] To be honest, if you had asked George Boole who was doing the most impractical research, he probably would have said he was as well. But Boole persisted, and just under a century later, the first computer was built – a device that was perfectly designed to utilize Boole's creation. All computer systems and all computer programs are built using his creation. Boolean logic is mandatory learning for computer scientists. Philosophy for philosophy's sake changed the world – it just took 100 years to do it.

In a stroke of cosmic irony, Boole's great-great-grandson took a similar path. Geoffrey Hinton was not interested in AI as a programmer. Instead, he was just interested in the human brain. Hinton obtained his bachelor's degree in experimental psychology, like so many people in this story, from Cambridge.[2] He was fascinated by the way the human brain functioned, and true to the title of "experimental," he became convinced that the key to understanding the human brain was not to be found in human experiments but in machines. With his bachelor's degree in hand, Hinton left Cambridge and ventured to the University of Edinburgh to study artificial intelligence.[3]

Although his program was specifically focused on AI, his teachers were not the mainstream computer science researchers we have been discussing thus far.

His PhD supervisor, Christopher Longuet-Higgins, was a famed theoretical chemist.[4] But, as is common with academics, he found himself getting bored of the research for which he was most famous. You see, most great academics join the academy because they love learning and problem-solving. In your early days at university, every problem is a nail and you are a hammer. There is so much to learn, and the process is exhilarating. However, as you progress through your career, you hyper-specialize. Within your area, you slowly learn all the major information there is to know. If you are talented, and lucky enough, you become a thought leader and are considered to be more knowledgeable than anyone else on the subject. It is a humbling accomplishment, but intellectually lonely.

However, Longuet-Higgins found a new problem to excite him – the problem of the mind. First as a curiosity, then as a profession, Longuet-Higgins dug into the complexities of the human brain and, eventually, the emerging field of artificial intelligence.[5] Alongside two other researchers – a neuropsychologist and a Bletchley Park codebreaker – Longuet-Higgins founded the machine intelligence program at Edinburgh.[6] This group would take a different approach from the mainstream researchers. While the mainstream focused on building "practical" systems and championed the symbolic approach, these researchers were infatuated with the way the brain operated. They wrote theories about how we naturally perceive and understand the world around us, and then tried to translate these theories to machines.

It was at Edinburgh that Hinton entrenched his belief in the neural network approach.[7] He thought that AI systems should mimic nature. When we learn, we are not given a logical set of rules to follow. We learn through experience. If machines were to be intelligent, they must learn through experience as well. All we had to do was build the right structure, and the machine would do all the "hard" work of learning for us. In the face of the whole mainstream telling him he was wrong, Hinton never wavered. He was convinced that, one day, neural networks would become the foundations for understanding the human brain and creating intelligent machines.

Although Edinburgh was a welcoming environment for "different ideas" about AI, Hinton's approach was still not fully accepted. His supervisor would constantly beg him to change his approach.[8] Even if such a program could theoretically be built (which seemed unlikely), there wasn't a computer in the world that would be powerful enough to run it. But Hinton remained unphased. Every plea from his supervisor was met with the same response: "Give me another six months, and I'll prove to you it works."[9]

Fortunately for Hinton, his adviser granted him the six months. Unfortunately, it took him over thirty years.

After getting his PhD from Edinburgh, Hinton found himself bouncing between universities as he struggled to secure stable funding for his work. As we discussed earlier, AI went through many winters and springs during this period.[10] Waves of excitement regarding the advancements of the mainstream meant that some money would flow to him, but the winters were sub-Arctic. In 1980, he moved to the University of Toronto, which would become the long-term home for his research.* It was in Toronto that he pioneered his work on deep neural networks, a technology that would change the world as we knew it.[11]

---

* The 1980s saw the neural networks approach experience something of a resurgence, pushing back against the dominance of the symbolic approach. This resurgence was driven by Bell Laboratories, led by Yann LeCune, and made massive strides in demonstrating the commercial effectiveness of neural networks for tasks such as handwriting recognition.

On the surface, deep neural networks are not much more complicated than the neural networks we have discussed thus far. However, anyone who has tried to build one will laugh – or more likely cry – upon reading that statement. They are VERY mathematically complex. Luckily for you, we will not be going into the math (partly because I don't fully understand it). Returning to Rosenblatt's Perceptron, we can observe a traditional neural network. In a traditional network, there is an input (the image), a single layer of nodes (the hidden layer), and then the output.[12] The image goes in; it is processed by the weighted nodes that have been previously trained by trial and error; and an output comes out.

A deep neural network is exactly what it sounds like: a neural network that is deeper. Instead of a single layer of hidden nodes, it has multiple layers. Such an architecture would allow for a vastly more complex analysis of data. Each layer gives the possibility of learning more information. Each layer can tackle a different part of the problem that the system is designed to solve.

But, of course, more layers meant more complexity, requiring a more complex program to operate the system and exponentially more computing power.

In 2004, after years of research laying the foundations for the development of deep neural networks, Hinton, alongside future AI legends Yann LeCune and Yoshua Bengio,† received a grant to create a new program exploring neural networks.[13] The "Neural Computation and Adaptive Perception" program would play host to and be a pivotal developmental space for many of the biggest names in AI right now.[14]

---

† I wish I had the time and space in this story to discuss these other AI legends and their contributions to the field of AI and deep learning in depth.

Things seemed to be going pretty well for the neural network crowd. They were still the outcasts of the AI community, but they were surviving, and sometimes that's enough.

On the other hand, the mainstream AI community found itself on unfavorable terrain. By the early 2000s, several AI winters had come and passed. Small breakthroughs would occur, leading to a wave of initial hype, followed by more bold predictions. Funding would come in, but no substantive progress would be made.[15] This pattern inevitably led to general disappointment and mockery of the field as a whole. During the last major wave, symbolic-based AI programs had been sold to businesses with grand promises from AI experts that never materialized.[16] Reality was starting to set in; it seemed that maybe the mainstream approach wasn't the way forward – maybe they had blundered.

Knight takes rook. The chess board was shifting, and the tides were turning. Hinton and his colleagues were growing ever closer, and the more they worked, the more it became clear. They were right. They were always right. But frustratingly, they couldn't deliver the devastating blow for which they were searching. The finishing move was starting to become visible, but they were far from the end game.

As we know now, the pieces did fall in Hinton's favor. However, the rest of the middle game was relatively out of Hinton's hands. He needed outside help. He needed more pieces to develop.

Sometimes great ideas emerge at the wrong time. When George Boole invented his logic, he had to wait 100 years before the computer was invented. Sometimes our minds outthink the limitations of the technology of the moment. Deep learning is an example of this. The theory was sound, but there were two practical problems. First, Hinton's PhD supervisor was right – computers just weren't powerful enough to run the programs he was imagining. And second, there wasn't enough data.[17]

For a deep learning system to work, it needs *a lot* of data. To train a system to recognize trees would require thousands of labeled photos of trees. Creating a system to drive a car would require billions of data points. For deep learning to be a tool that could actually be practical in the real world, we were going to need to produce a lot more data.

Luckily for Hinton, he wasn't going to need to wait for 100 years. Quietly, and without many people knowing about it, a revolution was occurring. The way humans interacted with the world around them was changing. A new period of human history was defining itself. And these changes delivered everything the supporters of neural networks needed.

Welcome to the digital age ...

CHAPTER 9

The Machine

As I open my laptop to begin writing this chapter, I am on a train going from Amsterdam to Paris. But this is no ordinary train ride. The conductor has just made a terrifying announcement.

"I regret to inform you that our wi-fi system will not be functional today," he declared.

Based on the reaction, you would have thought he told us there was an Ebola outbreak on the train. Yet even with this inconvenience, my surroundings are still awash with the glow of laptops and smartphones. In fact, as I look down the aisle now, I do not see a single person not using a computing device. The lack of wi-fi only phased us for a second. Then we all remembered – we have data.

I feel as if I often take for granted how unusual this situation is. It's not just the act of escaping into our devices, nor our deep dependency on them, nor even how interconnected we have become, but how quickly we seem to have got here.

In 1943, the president of IBM supposedly said, "I think there's a world market for maybe five computers."[1] To be fair, computers hadn't really been invented yet, and he had no way of knowing just how useful they would be. But fast forward to 1977, and we hear Ken Olsen, founder of a major computing company, state, "There is no reason anyone would want a computer in their home."[2] There were digital computers at this point, and there were even some personal computers. We knew their capabilities, and yet the future was still clouded and unclear.

Of the illusions we create in our lives, the one that fascinates me more than any other is our tendency to observe history as a static line. We often fail to consider how different things could have been – how our minor choices affect the timeline. If the right person hadn't been in the right room at the right time, if fate had not aligned just so, the world would be a different place. It can be a terrifying prospect. Many of our actions probably have no impact on the grand symphony of the universe, but looking back, those who do make history often do not know it. The smallest actions can often change the destiny of millions. The fabric of history's narrative is made up of billions of isolated decisions. Each decision thread comes together to produce the patterns of the society we live in.

So let me tell you another story – a story made of decisions, some foolish and some bold. The main character of this story is not human, like the stories in the first act, but the machine itself. It is not the story of how we got the computer of today, but how we came to interact with it – and are slowly becoming part of it.

MAKING THE MACHINE

Both of my parents are computer scientists. In my youth, they would tell me stories of the early computers they worked on. There were colossal machines that took up entire rooms. If you wanted to use one, you had to book time with it. To program the computer, they used punch cards, just as Lovelace had once envisioned.[3] The process was slow and not very user friendly. You quite literally needed a degree to understand how to use them.

They both went to school during the early 1980s and were living through a massive shift in the way humans and machines interacted. What they were engaging with were mainframe computers – large, bulky, not very user friendly, but damn powerful for their time. Not every mainframe computer used punch cards; there was now a more efficient way to communicate with the beast that was sometimes available: the keyboard.[4]

Ripped from the typewriter and connected to the computer, the keyboard allowed for far more dynamic conversations with the device. This innovation was significant as talking to computers is as difficult as it is important. It doesn't matter what capabilities a computer has if we can't tell it what to do!

Remember, what computers do is run algorithms designed by humans. We need a way to give them these algorithms and the inputs. Since digital computers work through a series of switches in a state of "off" or "on," the simplest way to communicate with them is binary – a string of 0s and 1s signifying on and off. However, in keeping with the theme of this book, simple is not always practical. To program in binary is excruciating if not impossible. That is why we invented computer languages.

Programming languages allow us to talk to machines with greater ease. We can write our instructions in a language closer to English, and a special piece of software will translate it into binary for the computer to execute. There is no universal computer language other than binary. Different languages such as C++ or Python provide different advantages through their structure and may be favored for different kinds of computing tasks. But all of them are different ways of requesting a task that will be translated to machine language.

By the early 1970s, personal computers had begun to materialize, but they lacked any real oomph and could cost as much as a house. Their designers were emboldened by the development of monitors to display information and the keyboard.[5] Although these developments were rudimentary, they were envisioned by some as the average person's portal into the computer.

The road to personal computing would not have been possible without another groundbreaking invention that lurked behind the screens: the microprocessor.[6] The early computers we have discussed, such as Babbage's engine, had a sole purpose – they were dedicated computing behemoths designed solely for calculations. Although more complex, early digital computers were the same. Microprocessors compressed the function of a computer into a smaller form. The Intel 4004 was the first one commercially produced.[7] It was tiny compared to mainframe computers, able to fit in the palm of your hand. And it could complete all the same functions, just not with as much power.[8] When you hear someone refer to a CPU, they are referring to your computer's microprocessor. It is the brain that makes your device function and the heart that pumps life into the cold hardware. With a microprocessor, a keyboard, and a display screen, the silhouette of the home computer was taking form.

Programming through a keyboard directly to a display screen was a game changer. It allowed programmers much more freedom and created the space for near endless innovation. But it did not open computing up for everyone. To use a computer, you still had to learn a totally new language. And although the keyboard was an upgrade, the computer still wasn't overly user friendly. The screens attached to the computer looked nothing like the ones we have today. The cutting-edge technology would simply be able to display the text the user was writing and the outputs of a program. With these drawbacks in mind, it's not difficult to understand why Ken Olsen didn't think people would want a computer in their home. What would a normal person without a computer science degree do with one?

By the late 1970s, the personal computing space began to heat up, and two new players entered the picture: Apple and Microsoft. Both were led by visionary leaders, Steve Jobs and Bill Gates, who dreamed of being the one to introduce computing into the home. Of course, the history of personal computing is much more complicated than just these two. But we will focus on them as their innovations and stories provide a convenient window into the developments that are important to our grander narrative. They are also undeniably the most important figures in this history, as their decisions shaped the computer's trajectory.

Steve Jobs is remembered for many things. His name summons images of the iPod, the iPhone, and his trademark turtleneck. Yet, one of his greatest additions to computer history is little discussed outside of computer nerds. A keen reader may have already noticed that there is one key part of the personal computer missing from our silhouette: the mouse.[9]

The story of the mouse is one of my favorites in the history of computing. Like the story of neural net researchers, it is a testament to true innovators' ability to see something that others cannot. The mouse was not invented by Steve Jobs; he took it from someone who didn't know what they had. No crime took place. It wasn't a stolen secret. It was a case of seeing a simple technology and realizing the future in it that its inventors couldn't.

When he found the mouse, Jobs was not yet wearing his turtleneck and was more commonly spotted in a dress shirt and bow tie. Yet, even without his signature outfit, everyone in Silicon Valley knew his name. He had soared to fame and fortune alongside his

Apple cofounder Steve Wozniak, spearheading the creation of two user-friendly home computers: the Apple I and Apple II.[10] The Apple II even had a keyboard, a text display screen, and a microprocessor.[11] Although they sold well, the market was still limited. Jobs's next project was to be much more ambitious. He wanted to find a way to make the computer more accessible to everyone. But he was stumped. How could he make the average person understand what was going on with a computer?

The idea he was searching for was revealed when he reluctantly agreed to visit a team of some of the world's greatest computer engineers: the team at Xerox.

Yes. That Xerox ... the printer company.

Although Xerox has fallen far, the name Xerox still resonates in the computing world. Many of the most legendary researchers in Silicon Valley are alumni of Xerox PARC, the research and development center for Xerox. PARC attracted some of the best researchers from around the world, who worked to develop forward-thinking computing ideas. At the time, one of the most technologically advanced systems they had developed was a laser printer. But it wasn't the printer that caught Jobs's attention. In 1979, Apple facilitated his visit to Xerox PARC.[12] It was there he saw the Alto – Xerox's version of the desktop computer.[13]

When Jobs told the story, he recalled being shown several groundbreaking inventions that day. But he only recalls one: the Alto's graphical user interface, or GUI for short.[14]

It has been said that Jobs hadn't been overly interested in adding a graphical unit to the computer before that visit, and the project had mainly been pushed by other Apple staff. But one thing is for certain: after Jobs saw the Alto, he was sold.[15] The GUI created a more natural environment with icons and menus. It could allow a complete amateur to interact with a computer with minimal to no previous training. And this miracle was locked away inside this research center. He had to free it. And more … what was that thing attached to the monitor? A small plastic brick attached by a wire. *It was the mouse.*

It was all coming together now. The mouse was the most perfect invention. It could act naturally as an extension of the user's hand. Just as the keyboard allowed us to extend our use of language to the screen, the mouse would allow us to extend our human touch. So natural, so elegant: just point and click.

If it hadn't been for Jobs recognizing the importance of the mouse, the computer of today and the way we interact with it might have been completely different. It is possible that personal computing may have even failed to develop a market. But Jobs's reluctant visit to Xerox PARC marks a changing point in history. Apple's next computer, the "Lisa," was to have a GUI and a mouse.[16]

The Lisa was a technological marvel. It is considered the first mass-marketed computer with a graphical user interface and the first with a mouse – a monumental step forward for computing.[17] But when the visionary future of personal computing hit the shelves in January 1983, it did not enter with a bang.[18] Instead, it was more of a pitiful whimper.

Being a technological marvel came at a price, $16,995 per unit to be exact. That's over $53,000 today.[19] In hindsight, it's not surprising that the Lisa was a massive failure. It failed to recoup its budget, and although it didn't mortally wound Apple, it put the company in a precarious financial position. Toward the end of the Lisa's development, Jobs himself realized the project's fatal flaw. Luckily for Apple, they had another project already in the works: the Macintosh.

The Mac had many of the same ambitions as the Lisa, and it's possible that, if it hadn't been in the works since the mid-1970s, long before the Lisa's failure, Apple might have scrapped the project. The primary difference between the two was that, instead of being marketed toward businesses who could potentially pay a high price tag, the Mac was envisioned as being cheaper, appealing to students and the general public.[20] Before the Lisa was finished, Jobs jumped ship to the Macintosh project, which had originally been the brainchild of Jef Raskin.[21] With him, Jobs brought over many of the ideas for the user interface that had made him fall in love with the Lisa.

Jobs's vision clashed heavily with Raskin's, causing Raskin to eventually leave the Macintosh project. Many accounts say that the Macintosh the world now knows is a distant cry from Raskin's original vision and that Jobs's vision for the future of computing dominated the finished device. Another divergence in the timeline. What might have happened if Jobs had never joined the Mac team, if we had gotten Raskin's vision instead? What might have the computer looked like?

When it launched in 1984, the Mac received a very different reception from the Lisa.[22] Although it wasn't perfect and had serious design flaws, it presented a romantic idea of what computing could be. Users across the globe fell in love with the device. Jobs had nailed his vision. He had created the new standard for personal computing.

Discussing the Macintosh, Bill Gates once said, "To create a new standard takes something that's not just a little bit different. It takes something that's really new and captures people's imaginations. Macintosh meets that standard."[23]

Even though it fell short of performance expectations, and sales slowed, the Macintosh sketched a clear image for what a personal computer was to look like.[24] People now knew what to expect from a computer that sat on their desk. Apple built on the original Mac, fixing its problems and propelling it to being the dominant player in the personal computing space. But Jobs was not around to enjoy the company's success.[25]

After the failure of the Lisa, Jobs brought in a new CEO, John Sculley, a marketing expert. Sculley is credited with the marketing campaign that helped cement the Macintosh's place in the public imagination. But over time, he became convinced that Jobs's eccentricities were a hindrance to the company's success and slowly pushed him out. In 1985, Jobs left Apple.[26]

He was not finished though. True to his reputation as a stubborn genius, Jobs patiently bided his time. When he did finally return to Apple, he would revolutionize personal computing for the second time in his life.

Bill Gates took a different path to computer stardom. Instead of focusing on the hardware, he directed his attention to the software.[27] The purpose of his company was straightforward: they would create software that could be run on the increasingly popular computer. Specifically, they would focus on microcomputers. Software for microcomputers ... Micro ... Soft. To be completely honest, I just learned that is where the name "Microsoft" comes from, and I felt quite foolish for never having noticed.

Designing software came with several advantages over Jobs's approach. Although the initial investment was high, as one had to program the software itself, the long-term costs were low. Even after the Macintosh was invented, Apple had to spend money to actually produce the physical computer. Microsoft, on the other hand, could create as many copies of their software as they wanted at almost no additional cost. This model, paired with ingenious design, quickly propelled Gates to the status of the world's youngest billionaire.

Microsoft's flagship product was their operating system. When Jobs introduced the world to the Macintosh, the standard personal computer form was born. Gates followed in Jobs's footsteps, designing a graphical user interface that could be run on non-Apple products.[28] Companies would pay small fortunes to secure rights to have their products shipped to the public with Microsoft Windows pre-installed.

Microsoft developed into a company that dominated the computing space. Their dominance was in part because of their quality products but also due to the fact that Gates made ruthless, and extremely effective, business decisions. One such decision was creating an alliance with Intel, the world's leading chip manufacturer. Recall, it was Intel's invention of the 4004 microprocessor that really kicked off the world of personal computing.[29] Such a relationship would be key, as Intel's innovation did not halt with the 4004. In fact, every year, they seemed to be able to put out exponentially more powerful microchips. In 1975, Intel CEO Gordon Moore predicted that the microchips they produced would double in power every two years.* It was an extremely bold prediction. Yet magically, it was right. Not only was he right for the foreseeable future, but he's been right to this day. His prediction has become known as Moore's law.[30]

The Intel-Microsoft partnership began in the early days of personal computing, not long after Moore first stated his law. The catalyst was IBM deciding to use both companies' technology in their microcomputers.[31] The two companies quickly realized that they worked well together and could collude to boost profit. There was no formal contract that defined the relationship, but the two worked in tandem. They communicated strategically and ensured that their respective technologies complemented and enhanced each other's performance. This partnership was not a secret agreement either, as the two ran conjoined marketing campaigns promoting the use of Intel processors and Microsoft software as the optimum way to build and run a computer.

---

* The actual prediction is that we will be able to double the number of transistors in an integrated circuit per chip every two years – which will double its power.

The plan worked splendidly, and "Wintel" became the standard for personal computing.[32] Every computer I owned before starting university was a Wintel computer.

At the same time as Microsoft and Intel were reigning supreme, Apple was fading into oblivion. They seemed directionless without Jobs, and the products they launched seemed uninspired, or worse, useless. Further, they lacked the usability and adaptability of Wintel devices. Although Apple had their loyalists, Wintel devices were the standard. They were more affordable and had a much larger ecosystem of software that could be run on them. Although Apple defined what the personal computer would look like, they failed to be the ones to benefit from its rise. This wasn't a mistake they would make twice.

In 1997, Apple welcomed back Steve Jobs as CEO.[33] He again proved his visionary prowess. In ten short years, he managed to revitalize the company and created a revolutionary device that would usher in the end of Wintel dominance.

But during Jobs's exile, another revolution had been brewing: a revolution that is potentially more important to our story than any other – the internet.

CHAPTER 10

Ties That Bind

My experience on the train shocked me, as it revealed just how uncommon it is not to be connected to the internet. It's all around us. And it is truly marvelous. The collective knowledge of humanity, openly accessible to anyone. Wikipedia alone contains more information than the Library of Alexandria.[1] If I have a question about Alan Turing's work, it only takes a flick of my finger and I can read countless articles chronicling his life.

When I am traveling, I attend meetings through Zoom. My face and voice are projected effortlessly to my colleagues thousands of miles away. When I'm bored, I watch YouTube and Netflix. If my favorite hockey team is playing, I stream the game live from my phone. Although my friends are scattered around the world, I can track them through social media and keep in contact as if they were living in the same city as me. The internet is magical. But it can also be a terrifying and destructive place.

Despite its colossal presence in our lives, the internet comes from humble beginnings. It was never meant to become the behemoth it is now.

The original idea was quite logical. In the aftermath of the Second World War, more and more computers were popping up around the world. At the time, the computers were large and mainly confined to research labs and universities.

"There aren't that many of them, so why don't we connect them?" the researchers thought. "Then we can share our data and findings faster!"

"A marvelous idea!" thought the American government.[2] "This can help ensure that American innovation happens faster than Soviet innovation."

Never underestimate the American government's will to fund science when they want to stick it to the Russians.

The project of attaching the existing computers was called the Advanced Research Projects Agency Network, or ARPANET.[3] Although it was not without its technical challenges, ARPANET benefited from existing infrastructure that made its mission much easier: specifically, telephone lines.[4]

Why invest in new cables to stretch across the country when the telephone companies have already done it? Within a couple years, the researchers figured out most of the technical problems and managed to connect their computers.[5] The internet was here. Hurrah!

In reality, what they had created had little resemblance to the internet we are used to. All it could do was send data back and forth between the limited number of computers on the network. But it lit the fuse and laid the groundwork for the system we now enjoy. I think ARPANET is also a blunt reminder of what the internet actually is. We are so often distracted by the memes and flashiness of the internet that it's easy to lose sight of what it's actually doing. In reality, the internet is just a system to send data from one computer to another.

When I write an email, my message is encoded into data and jettisoned through the internet to my friend's computer. When I watch a YouTube video, YouTube's computers are sending me that video in the form of data. The internet is the medium through which data is transferred around the world.

As computers evolved, so did the internet. Yet for the majority of its short history, the internet was a walled garden, only primarily accessible by large government-funded institutions.[6] It wasn't accessible to everyday people. To be fair, although the internet was pretty cool, everyday people weren't missing out on much. Back in the 1980s, when personal computers started getting popular, the internet was still just a means of sharing data between computers. There were no websites. And the early internet was painstakingly slow.

By the late 1980s, ARPANET had started to become outdated, and its funders were ready to move on. So, they handed the job of looking after the network of computers to a new group, the National Science Foundation (NSF), which launched NSFNET.[7] For a while, NSFNET kept the internet just for research and education, not business. But in 1989, they started changing the rules, letting some business in.[8] Around this time, companies that provided internet access, called internet service providers, began to pop up.[9] They started selling internet access to everyday people and businesses.[10] Shortly after they entered the picture, NSFNET shut their doors and ceded control over the internet to the internet service providers.[11] The internet had been freed! Or so we thought.

As control of the internet traded hands one last time, a new version of what it could be exploded into existence: websites.

Navigating the internet before the invention of websites was a lot like navigating the highway system before the invention of GPS. You had to know where you wanted to go, and heaven forbid someone had closed a road, or your favorite pizza place had changed locations without telling you. Receiving and sending data was a frustrating science. And it was messy. Information constantly got lost in nooks and crannies of the internet ... there had to be a better way.

"What if we just connect all the data with hyperlinks?" a voice called out.

"What did that man just say?!?" the world responded.

In 1989, Tim Berners-Lee proposed the idea of the World Wide Web.[12] Instead of maps and roads, every internet user would have a personal teleporter that could take them wherever they wanted on the internet.

Berners-Lee was inspired by the failures of the internet. At the time, he worked for CERN, a massively important research center in Europe.[13] He proposed to his bosses that they could create a system, built from a patchwork of already existing technology, that would make the internet a million times easier to use for research.[14]

The plan centered on hypertext.[15] When you copy and paste a URL and it appears underlined in blue, that is hypertext. It was not a new invention; it was commonly used to link together multiple documents that were stored on a single computer. If I had a document for a business pitch that referenced a financial report, I could add a hypertext link from the pitch document to the financial report. Berners-Lee's idea was to apply this technology to the internet.

Every page in the web would have a unique hypertext name, called a universal resource locator, or URL for short.[16] Pages, or "websites," could hyperlink to other related websites using hypertext and the other websites' URLs. To access these sites, the user needed a special program called a web browser. If I clicked on a link, the web browser would search for the data associated with the URL and beam it to my computer. Additionally, if you knew exactly where you wanted to go, you could type the URL directly into the web browser, and it would find the data for you and display the website.

The World Wide Web proved to be extremely useful at CERN. In 1991, Berners-Lee invited the world to enjoy his creation, making the first web browser public and sharing the instructions for how to create a website.[17] Just like that, the World Wide Web sprang into existence, and the digital age truly began.

When discussing AI and its relationship to the internet, it is important to maintain focus on the fact that the internet is made of data. When we go to a website, it is not conjured out of thin air on our computers. It is data that exists on a computer somewhere else in the world. When we visit a website, we are requesting that this data be sent to our computer.

The years following Berners-Lee's invitation to join the World Wide Web were an explosion of innovation and failure. The internet was about to go through a massive transformation. New web browsers such as Mosaic (1993), NetScape (1994), and Internet Explorer (1995) entered the picture, providing access to the World Wide Web to the masses.[18] At the start of 1993, there were only 130 websites. By 1996, the number had exploded to over 100,000, and websites such as Amazon and Yahoo had been founded.[19] By the end of 1997, there were over 1 million websites, including Netflix (although not quite the Netflix we know today).[20]

"There's a lot of money to be made on the internet!!!" screamed the world's young and hungry entrepreneurs.

"This is so cool! You must be able to make so much money here!" investors agreed as they danced with foolish glee.

Thousands of new websites emerged. The internet became a hotbed for online retailers and every paid service imaginable.

Trillions of dollars were hastily thrown at internet-based companies who had no real plan for how to turn a profit. What could go wrong?

A lot.

There's an economic myth that the market is smart and always knows best. But that myth ignores the fact that the market is not a single godlike being. The market is constructed out of the decisions of millions of people. And people are often dumb.

Eventually the free money train, also known as "investment," stopped, and the internet companies were expected to be profitable. But they weren't. They were actually highly unprofitable.

This was the dotcom bubble.[21] Thousands of companies, thought to be worth billions of dollars, were completely unviable. Their "worth," as determined by the market, was just a dream. The companies had no way of producing actual profit through their services. The "value" that appeared on a dotcom company's stock ticker was not a representation of its actual worth but rather an illusion produced by hype and excitement.

Economic bubbles are intoxicating. They prey on our human desire to believe. When you're in one, it seems like it will never end. But when they pop – when those involved realize the value was never real – boy do they pop.

The dotcom bubble peaked in March 2000. By 2002, investors were estimated to have lost $5 trillion in the crash.[22] The internet economy was devastated.

But why is this crash important to our story and to Geoffrey Hinton in particular? Because we can draw a direct line from the dotcom bust to the AI landscape of today.

Those who emerged from the trenches of the dotcom bubble were battle hardened. The companies that remained knew they were never going to be given "free money" again. One company that scrapped its way out of the collapse of the bubble was a small search engine named Google.

Founded in 1998 by two Stanford PhD students, Larry Page and Sergey Brin, Google offered a much more user-friendly method for surfing the web.[23] Google was not the first search engine, but it excelled where others had fallen short. Search engines needed to scrape the web to determine what was on each website. But just knowing what was on each site wasn't enough. If a user searched for the word "cars," there would be thousands of relevant pages. The secret sauce was determining which pages should be shown to the searcher. There needed to be a ranking. To achieve this, Page and Brin designed an ingenious algorithm that ranked pages based on how many times other pages linked to them.[24] Looking back now, the method was quite rudimentary compared to the page rank optimization we enjoy now. But at the time, it was revolutionary. There was a problem though: how were they going to make money off it?

Surviving the crash took financial prudence. You could not burn through cash or rely on potential future profits. Companies needed to make money in the short term. There may be an alternative timeline in which Google picked a different business model – perhaps a payment for service model, where each of us paid a $5 monthly fee to use Google. But no. Google chose a business model that would reshape the online economy. Google Search would not be their product. Instead, it would become a platform for their product: advertising.[25]

In the year 2000, Google introduced AdWords.[26] The original system was not groundbreaking; it just mirrored the advertising model of newspapers, admittedly with some technological upgrades. Advertisers paid to have their ads shown when users searched for specific words.[27] The amount they paid was based on how prominent they wanted their ad to be and how many impressions (views) they wanted their ad to have. Google's advantage in marketing stemmed from its precision. Unlike traditional methods, Google's digital platform ensured advertisers could accurately track views, guaranteeing that their ads reached the precise audience they had paid for.[28] Contrast this precision to a newspaper, where there was no way to know exactly how many people actually looked at the page you had paid to put an ad on. Moreover, if you were selling a product like cars, you could ensure that your ads would be shown to people searching for cars. It was this precision that excited advertisers and made Google a financial titan. Google had a winning formula, but they were not yet satisfied.

In 2002, Google shocked the advertising world by shaking up their advertising system.[29] No longer would advertisers need to pay per impression. Instead, they could choose to "pay-per-click."[30] In normal speak, they would only need to pay when someone clicked on their ad.[31] Although this shift may seem small, it was paradigm altering. Suddenly Google took on immense risk. They could end up showing thousands of ads to users without making a cent. But Google had a plan. The switch to pay-per-click came with a trade-off. They would no longer be positioning ads on a website solely based on who paid the most money. Instead, ads would be positioned based on which ads Google thought the user was most likely to click on.[32] Google had created an algorithm to attempt to predict the desires of the user based on the keywords in their search.[33] In 2009, building on this strategy, Google began leveraging users' individual and historic data to better predict their behavior.[34] The user's search history became a variable, fed into an algorithm, to achieve a goal: this approach allowed Google to present an advertisement that users were likely to click on.

Google eventually launched another revolutionary advertising technology: live auction.[35] In a live auction, companies interested in advertising their product describe what kind of person they want their product to be shown to. They may provide keywords or use more advanced consumer analytic strategies.[36]

Now, imagine you just did a Google search. There is a fraction of a second between the moment you hit search and the moment the result page displays. In that split second, electronic mayhem occurs. Google instantly generates a profile of you in relation to the search you just made. They then pitch that profile to thousands of advertisers who have a stake in the live auction. Then, a massive algorithm-fueled bidding war occurs. The advertisers' algorithms bid on the right to advertise to you according to how valuable your eyeballs are to them, based on the profile Google has provided.[37] When the dust settles, your attention has been sold, and a page appears with tens of paid ads tailored just for you – all in the split second it takes to load the page.

The money couldn't stop rolling in, and Google dedicated themselves to perfecting the juggernaut they had created. They had transformed the static classified advertising page into a dynamic and personal experience. Again, the secret sauce was personalization. To innovate, Google needed to continuously enhance their ability to match advertisements with users.[38] It was the beginning of what is now known as "surveillance advertising."[39] Almost overnight, the internet was transformed into a massive surveillance machine.

The thing we call data is the product of surveillance. We often think of it as being something digital, but it's not. If you went to a baseball game and recorded every pitch with a pencil and scorecard, you would be creating data.

At its most basic level, data is a collection of perceived facts. Your baseball scorecard is a collection of perceived facts about the outcome of each pitch. It is perceived because sometimes there are mistakes. Maybe you recorded a ball as a strike. It does not change the fact that you have produced data. Bad data is still data. And all data, good and bad, would become the bedrock of AI.

To create data you need three things: a subject, a watcher, and a desire to create data.

Every subject in the world, whether it's you right now or a rock on Mars, has the potential to create data. But to become data, someone must decide that it should be data, and they must desire the event or action be recorded. Then they will decide how they are going to record it. They need some kind of technology, like a pencil and paper, a computer, or a camera. Then the watcher records the action of the subject on the medium, and voila! Data is created.

Data isn't new; it's been used for thousands of years to organize our societies. We use it to try to make sense of grand problems or situations: to make observations over time. But something changed in the digital age. Before that, data was more difficult to come by. Someone had to go out and collect it. But computers were constantly producing and storing data. Every word you type on a computer is inscribed and stored as data on the machine. With the invention of the internet, the amount of data produced every day skyrocketed. Every time we clicked on a link, data was produced. Every action had the potential to be surveilled and turned into data.

Google was potentially the first company to truly grasp the potential of all this data.

They began scooping up as much of it as possible. They designed systems such as Google Accounts so that they could link the data to individual users. They leveraged technologies such as third-party cookies that would allow them to track users across the internet and collect data on them, even when they weren't using Google-owned websites.[40] To be as tailored as possible, they needed more data. The secret: build and release free products that people would happily use, blissfully unaware of the data they were giving up. They launched Google Maps, bought YouTube, and in 2008, they launched Google Chrome, their own web browser, giving them an even greater eye into the web-surfing habits of their users.[41]

And what did they do with all this data? They used it to sell advertising – to try to predict your future behavior and sell it to the highest bidder.[42] It is likely that Google knows more about you – more about your darkest secrets – than even your closest friend.[43]

To sell their ads, Google became the world leader in data analytics. They produced immensely powerful algorithms that could predict users' behavior.[44] Naturally, this process led them down the path of artificial intelligence.[45] Like the rest of the industry, they didn't see much promise in the work of Hinton or the other disciples of neural networks. However, they did take a keen interest in machine learning.[46] They quickly became one of the most forward-thinking companies in the area, leveraging their data to create adaptive algorithms.[47] But they didn't just use it to enhance their advertising; they also realized they could use it to improve their public-facing services like Search and Google Maps.[48] Machine learning is the reason why the YouTube algorithm is so successful. It leverages the data of your viewing habits (how long you watched a video, whether you liked it, whether you commented) and calculates what content you are likely to enjoy.

Google may have been the first, but they were not the only company to leverage the data revolution. Thousands of smaller companies emerged that specialized in processing and making sense of user data. Data brokers who specialized in obtaining and selling data made their first appearance as well. But they were all picking up scraps. None of them could rival Google. Then, in 2004, a new website appeared: Facebook.[49]

Facebook took the Google model to new heights. Instead of sneakily finding ways to get information from its users, Facebook just asked.[50] Are you in a relationship? Who are your friends? What do you like? The age of social media had arrived.

Suddenly, interactions that had once taken place in public transitioned to a single place on the internet. We began organizing parties and building communities through Facebook. In high school, we had a Facebook page for every sports team or event where we would plan our activities. And Facebook was there to turn all of it into data.

The economy of the internet was simple – get data and leverage it into advertising. And the best way to get data? Find a way to transition an activity that usually happened in the real world onto the internet, where surveillance is easy. Little did we know that a new device was about to emerge that would make that task easier than ever, that would connect humans with the digital in a way never before imagined.

"Did you see what Steve Jobs announced yesterday?" my friend once asked me during recess.

At the time, I didn't even know who Steve Jobs was.

CHAPTER 11

The New Machine

If you were going to die soon, would you want to know? How would you react to the ticking clock of mortality?

When people are told they only have a short time left to live, there are two common reactions. One, they crumble under the weight of the time they have remaining. Two, they get to work.

Steve Jobs got to work.

In 2003, Jobs was diagnosed with a rare form of cancer and told that his days were numbered.[1] This was a very different Steve Jobs from the one we last visited. Gone were the bow ties. Now he wore the same sleek black turtleneck every day. He was focused and more driven than ever. And now he had a deadline. He had one last chance to change the world.

When we last saw Jobs, he had been ousted from Apple, an exile that lasted eleven years. But Jobs didn't just twiddle his thumbs for those years. He had many accomplishments while away from Apple. He started a new tech development company called NeXT, and in 1986, he paid $10 million to become majority shareholder and chair of the board for a small computer graphics company called Pixar.[2] I don't think I need to tell you what happened next with that company.

While Jobs seemed perfectly fine without Apple, Apple was not doing great without him.

The company lacked focus and launched failed product after failed product. Their boardroom had become a carousel. CEOs came and went, each unable to right the ship. They needed a visionary leader again.[3] So, in 1997, they purchased Steve Jobs's new company NeXT.[4] Quickly after, Jobs became Apple's new CEO.[5]

His return to the company was ruthless. When he parted, he had left behind a revolutionary new computer. Apple had been at the forefront of technology. But what he found was an uninspired, chaotic workspace. A heavy hand was needed to save his brain child from disaster.

Jobs wasted no time, coldly cutting numerous projects that did not fit within the scope of his vision. When the dust settled, he had cut 70 percent of Apple's product line, reducing the company's focus to just their computer.[6] Projects were not the only things at risk during Jobs's return. In a cold yet calculated move, he hastily fired 4,100 employees.[7] In one year under his leadership, the company went from losing over a billion dollars to making a healthy profit.[8]

In Jobs's mind, these dramatic measures were necessary. The company had lost its way and needed to be stripped down to its bones in order to be rebuilt.

The beauty, and danger, of Steve Jobs was that he was not an engineer. Instead, he sat somewhere between the public and the technology. He had an innate ability to look at emerging technology and ask, "How can this best serve the people?" instead of "What are the coolest things we can do with this?" His goal was to make computers that people wanted to use and enjoyed using. And that's what Apple did again after he returned.

As the company settled into its new rhythm, Jobs decided it was time to branch out past computers. Portable MP3 players were the logical next step. Or at least to Jobs it was logical.[9] They were a consumer electronic device that had become popular. However, Jobs and his team noted that the existing products were not very user friendly. Therefore, they fit nicely within Jobs's vision. Apple would create a portable MP3 player with a focus on the user experience.

In 2001, Apple released the iPod, a heightened MP3 player that took the market by storm.[10] But as the popularity of the iPod grew, Jobs sensed trouble on the horizon. Apple's finances were now tied to a device that was destined to fail.[11]

Sure, everyone had an MP3 player at the time. But Jobs couldn't ignore the growing popularity of cell phones. Every day, people were leaving for work and putting their cell phone in one pocket and their iPods in the other. This situation just wasn't sustainable. Jobs saw the future clearly: eventually the cell phone would eat the iPod.[12] All it would take is one company to make a successful phone with a built-in MP3 player, and Apple could be toast. He needed to be the one to do it first.[13]

In this decision, the seeds were planted for his greatest gift to the world.

The company's first venture into the phone market was an unmitigated disaster.[14] They partnered with Motorola, then the king of cell phones, to make the "iTunes Phone."[15] But Motorola didn't let Apple have any say on the actual design of the phone; instead, Apple took a back seat, only designing an iTunes interface for the device.[16] The whole project went up in flames. When presenting the phone to the media on launch day, Steve Jobs couldn't even get the iTunes feature to work properly.[17]

This fiasco must have been extremely embarrassing for the immensely proud Jobs. But he had an ace up his sleeve. A year before the failed launch with Motorola, two years after he had been diagnosed with cancer, Jobs had summoned two of his top engineers to begin working on a top-secret project: a cell phone that Apple would have full control over.

Both engineers were to lead their own teams and produce a prototype phone.[18] One team, the hands-down favorite, was tasked with making an iPod phone. The basic idea was to take the popular design of the iPod and add phone capabilities to it.[19]

The second team was to explore a more ambitious path. Apple had been playing with touch technology for quite some time.* Although they had yet to make anything too practical, some of their demos had caught Jobs's eye.[20] The second team was to explore a totally new concept for a phone: a touch phone.

The touch phone team quickly had a game-changing revelation. Moore's law had been steadily marching forward, and computer chips had been getting exponentially more powerful. Not only that, but a new type of computer chip had been growing in popularity; it was known as an ARM processor.[21] ARM chips were unique because, although they were less powerful than traditional chips, they produced less heat,[22] which meant there was no need for a fan like the one you find in a normal computer, making them perfect for use in cell phones.

All this came together to make the touch screen team think, "What if we don't just build a cellphone? What if we build an entire pocket-sized computer?"

Their idea: take the existing operating system used on Apple computers and shrink it down to fit on a touch screen phone.[23] Jobs became obsessed. Here was the vision of the future he was looking for. This was his final curtain call.

---

* Apple did release a touch device called the Newton during Jobs's exile. However, it had required a stylus and used much more rudimentary touch technology than what they were now exploring. The Newton was one of the first projects that Jobs axed when he returned to Apple in 1997.

In 2007, after a grueling development process, the iPhone was launched, albeit with a bit of illusion and trickery.[24] You see, I told you that this story was one of decisions, some good and some questionable. When Steve Jobs launched the iPhone, he made a very questionable decision. Come launch day, the phone was not quite ready. By not quite ready, I mean it didn't work at all. But Jobs refused to push back the launch. Instead, the demo was a carefully choreographed dance. Jobs followed a specific path while showing off the phone that the team had got working ... most of the time.[25] If one little thing went wrong – if he mixed up the order or stayed on an app for a split second too long – the phone would crash.

The iPhone launch is now remembered as one of the most important moments in contemporary history. But it could have easily been a disaster, tanking the product before it even had a chance.

But as the engineers sat at the back of the audience, crossing their fingers and taking shots of whiskey, they witnessed a miracle.[26] The phone didn't crash. And the world became obsessed.

The mouse and graphical user interface had been game changers because of how they allowed humans to interact with the machine. A touch-enabled computer that could fit in our pockets, that could connect to the internet – this invention would fundamentally change the way we interacted with computers.

Before the iPhone, developers were constrained by the silhouette of the computer that Jobs brought into existence. First, computers were relatively stationary devices. Second, programs had to be designed to be interacted with through a fixed keyboard and the mouse.

But suddenly, that wouldn't be the case anymore. An application could take any form a developer dreamed up. Keyboards and control pads could morph and change to fit our needs of the moment. It was the perfect design for a pocket-sized computer and kickstarted the smartphone revolution.

In a flash, the computer went from something that sat on our desks to something we carried in our pockets. We are literally glued at the hip to our computers.

Not only did this development mean that we had more computing power available to us at any moment, but it also meant that the computer was to become more embedded in our everyday lives. More and more of what we did translated into data. The iPhone even had a built-in digital camera, providing the ability to save any moment in digital form.

The iPhone is only one invention, but it helped us mark a significant turn in human history. We had become digital beings. The computer had started as a behemoth device that only existed inside research centers and universities. But sneakily, it had crept out. Its takeover was subtle; it happened slowly, then all at once. Society was to be completely reordered around the computer, unknowingly producing more data than ever before and investing trillions of dollars to build faster and more powerful computers every year.

Yet, even with all the advances occurring in computing, no one was talking about AI. Its failed promises had been resigned to the history books. But all this time, Hinton had kept working, and if you recall, when we left Hinton, he was only missing two things: data and computing power. By 2012, that was no longer a problem.

CHAPTER 12

The End Game

Geoffrey Hinton and his team didn't just weather the middle game, they silently dominated it. Against all odds, their strategy of patience worked out. The pieces on the board developed in their favor, and as the world became more digital, more and more data became available. Moore's law methodically pushed the pawns down the board, one square at a time, creating unrelenting pressure on their opponent. This was the end game. Years of evolution and preparation had led to this moment. Now the team just needed to set up for their finishing move.

The opportunity to demonstrate the power of deep neural networks to the world would come in the form of a competition. At the center of the competition was a database called ImageNet: a massive database of digital images.[1]

We have talked a lot about the digital revolution. But thus far we have not discussed one of the most important movements within it – digital images.

THE DIGITAL IMAGE

I have very fond memories of my father's film camera. He was deeply proud of it and had paid a handsome sum for it. After he had kids, he wanted to make sure he had the best technology to capture our family's memories and to freeze them in time.

It looked similar to the DSLR cameras that every amateur photographer has today. But other than their general function, the two technologies couldn't have been more different. When my father wanted to freeze a memory in time, he would line up the shot and snap a photo, just like today's cameras. As he pressed the capture button, the camera's shutter would open, letting light enter the lens. The light would scream down the barrel until it hit a piece of film: a translucent strip lined with chemicals that react to light. In an instant, the moment was burned onto the film, and the image was physically captured. But images captured on film don't always look exactly how they do in real life. A film camera is not capturing the moment itself. Instead, it is capturing the reaction of light and chemicals, combined to create a product that resembles the moment in front of the camera. Different lenses we put on a camera play with how the light is captured and distort the way it is recorded on film. Different types of film will capture the image differently based on what type and quality of chemicals are used.

Once captured, there was no way for my father to look at the image. If you opened the camera to examine the film, you would destroy it. The chemicals would react to the light, and the image would be erased. Instead, he had to give the raw film to a professional who, through another complicated process, would use chemicals to transfer the light data from the film onto special paper. Days later the developed photos would be returned, and our family could relive the memories.

But then, seemingly out of nowhere, a new technology emerged: the digital camera. My family's first digital camera was a dinky little thing, but it seemed like a miracle. Not only was it as simple as point-and-shoot, but we could instantly view the photos on the display screen on the back of the camera. What was this sorcery?

The general idea was the same – a lens and shutter lets light into the camera – but the way the light is recorded is fundamentally different. Instead of a piece of film, the light hits a special electronic sensor. This sensor is made up of millions of tiny little sensors arranged in a grid that each record the intensity and color of light they are exposed to. Each sensor then records the intensity and color on the tiny piece of the grid they are responsible for as computer data – a string of 1s, and 0s. The 1s and 0s for the entire grid are then saved as a file in one long string. We call this a RAW file. These 1s and 0s are the equivalent of my father's film. They have recorded the light that entered the camera through the lens. However, instead of needing a specially trained expert to develop the photo, all you need is a special program.

When you view the photo on the screen on the back of the camera, the computer chip in the camera is taking the 1s and 0s and using them to recreate the image by coloring computer pixels on the display screen to match the sensor data. Together, the millions of colored pixels create the illusion of an image. Digital photos are just a recording of light, in the form of color, per section on a grid.

This process allows us to take the real world and display a mirror of it in digital form. It also has some interesting side effects. Remember that company Steve Jobs bought when he was kicked out of Apple? Pixar. Well, they did not start as an animated film company. Instead, they were a research group at George Lucas's company Lucasfilm.[2] Lucas was obsessed with digital technology and was the first person in the film world to recognize its transformative potential. Before the innovation in special effects driven by Lucas's vision, special effects were practical camera tricks. The spaceships in the original Star Wars were physical objects (miniatures) recorded onto film. But with digital images, you could theoretically edit an image itself. You could change the 0s and 1s to edit or add objects that never existed.

The Pixar team broke off from Lucasfilm because they became obsessed with creating completely digital worlds. They wanted to make computer-animated movies. Due to financial restraints and respect for the artistic vision of the engineers, Lucas sold Pixar to Jobs, who committed to funding the animated films. But the computer engineers who remained at Lucasfilm, or who joined later, dedicated themselves to finding ways to transfer film into digital formats and then alter the images themselves. One of Lucas's engineers, John Knoll, leveraged these skills to co-create a consumer product called Photoshop,[3] a system that allows someone to change the 0s and 1s in any digital image. Want to remove a zit? Just remove the 0s and 1s in the code that represent the zit! Meanwhile, pioneering engineers within Lucas's companies found ways to digitally create new images and to splice the 0s and 1s for the new images into the 0s and 1s of the original image. Thus started the age of CGI effects.[4]

Importantly, in order to process and create these images, a special type of computer chip is used known as a GPU, a graphics processing unit. These chips are damn important to our story.

In a normal computer chip (a CPU), computer code is run in one long chain.[5] The more powerful a CPU is, the faster it is able to run the complex instructions given to it. Although this system works fine for the brains of the computer, it's not very good for processing images.[6] When assembling an image, you need to process the information for each individual piece of the image. Doing each piece in order can take a long time and is a bit of a waste of a CPU's abilities, as each piece does not take much raw power to process. So we invented the GPU, a separate chip that, instead of doing one powerful calculation at a time, could do a bunch of smaller calculations all at once.[7]

Think of it this way. You want to paint your house. You meet this amazingly talented painter – let's call him da Vinci. He is the CPU. He tells you he can paint your house to be the most beautiful house in the world! But it will take him six years and cost you $30 million. You tell him, no, you don't want the most beautiful house in the world, you just want it to be blue. He says, in that case, it will take me a week to paint your whole house, and because you are hiring the greatest painter in the world, it will cost you $200,000.

This seems unreasonable for the task at hand, so you call your local house-painting company. They are the GPUs. They send twenty painters named Doug who paint the whole house in one day and charge you $2,000. None of the painters are as talented as da Vinci, but they don't need to be. They each just need to take their own small piece of the house – and paint it blue.

IMAGENET

So now computers had a way to display images, and we could manipulate them. But wouldn't it be cool if computers could also understand images like we do? If a computer could process an image and tell us what was in it?

The task of image recognition has always been one of the main goals of AI. It's what Rosenblatt's Perceptron was designed to do: recognize whether an image was a triangle or a square. Very early on, researchers realized that, if we could teach computers to recognize images, the practical applications would be revolutionary.

We could have image search tools, facial recognition systems, self-driving cars, instant text translation, autonomous weapons and targeting systems … The military and civilian potential for such a system made it extremely appealing to researchers, especially since the military has a lot of money for funding technology. That's why they funded Rosenblatt's Perceptron.

In 2006, Fei-Fei Li, an AI researcher at Stanford, started a database called ImageNet, hoping to speed up the development of image recognition software.[8] The database was open source, meaning anyone could use it, and contained millions of digital images.[9] What made these images special was that they had been labeled.[10] Each image had been hand-labeled with a category explaining exactly what the image was. An image of a banana would be labeled "banana." An image of me would be labeled "person." In theory, a successful image recognition system would be able to take the raw image and process it through an algorithm to produce the proper classification label.

Image recognition programs are algorithms that, given an image as a variable, produce the outcome of a category label.

The success of the algorithm can be determined based on whether it produces the same label as the human who labeled the same photo. Thinking back to Alan Turing's definition of intelligence, we are judging the algorithm's intelligence by its ability to mimic our ability to label a photo. If it creates the output we would expect from a human, it is a success.

In 2010, Fei-Fei Li and the ImageNet team started a yearly competition, challenging researchers to compete to see who could train the best image recognition system on ImageNet.[11] One of Geoffrey Hinton's PhD students, Alex Krizhevsky, decided he wanted to take the deep learning techniques he was learning from his work with Hinton and apply them to the challenge. With help from Hinton himself, and fellow student Ilya Sutskever, Krizhevsky would create a deep learning neural network called AlexNet.[12] AlexNet was not the first program of its kind. But like many success stories, timing was key. You can't just have the right stuff. You need to have the right stuff and be in the right place at the perfect time. AlexNet had all three.

The theory behind a deep learning neural network like AlexNet is quite simple in the abstract. The goal is to build an algorithm that teaches itself how to recognize images in a similar way to how a human might.

For example, if I had a young child and wanted to teach them the difference between different animals, I might take them to an animal shelter. The first cage we walk up to has a dog in it, and I tell the child, "This is a dog."

"Dog!" they repeat back.

Then we walk over to a cage with a cat in it. "What is this?" I ask.

"Dog!" they respond again.

"No, that's a cat."

Now that the child has been corrected, they make some rules in their head about the difference between a cat and a dog. One such rule is that a dog is big and a cat is small.

As we continue, I show them more cats and dogs, and they are able to successfully guess which animals are cats and which are dogs. But then I show them a Chihuahua. The child sees the dog's pointy ears and small size and declares confidently "CAT!," only to be corrected moments later.

The model was wrong; not all dogs are big. Dogs can also be small. We repeat this process all day, meeting many different animals. By the end of the day, the child can confidently identify most animals that were in the shelter. After we leave, they can go into the world and, without my help, recognize any of the animals we saw in the shelter.

I never told the child rules about how to classify animals. Instead, the child learned on their own with my guidance.

A deep learning image recognition system is similar. At first, when shown an image of a cat, the system will make a wild guess as to what classification it should be. It will then be shown the correct predetermined answer. Over time, and with enough trial and error, the system should be able to adjust its nodes until it can accurately guess the classification for an image.

Although simple in theory, the architecture to build such a model is immensely complex. Deep learning is used since, with each extra layer, a new level of pattern recognition complexity can be added.[13] While the first level of digital neurons may only be able to recognize the general shape of the subject in the image, the deeper layers, with information such as the shape provided to them, may be able to decipher more complex attributes such as whether the image depicts a living organism or if it has wings.

A vitally important learning technique applied by the AlexNet researchers was backpropagation.* This technique is a function that allows the system to check its answers and see where it went wrong.[14] If the system were given a photo of a car and labeled it as a horse, backpropagation would allow the system to calculate why it may have been incorrect and send a message back through the architecture, adjusting the weighting of the nodes in an attempt to fix the error.

AlexNet may look very different from al-Khwārizmī's algorithm, or even those designed by Lovelace or Turing, but at the base they are the same. They are all a set of instructions to achieve a goal. At the end of the day, it's just highly powerful pattern recognition software. It does not understand the difference between a real-life cat and a dog. Or what a dog actually

---

* The invention of backpropagation is often inaccurately credited to Geoffrey Hinton. However, even Hinton himself admits the technique existed long before he used it, and he has never personally claimed to be the one to invent it. However, Hinton is one of the major figures responsible for reviving the technique and recognizing its importance to deep learning.

is. However, given a collection of data points that construct an image of a cat or a dog, it can recognize patterns in the 1s and 0s that provide it with an answer.

The major difference between AlexNet and other algorithms is that it is not static. It is fluid and ever changing. It is an algorithm that learns from its mistakes and can try and adjust to better achieve its goal. It may even be called artificially intelligent.

The deeper the architecture, the more accurate the system can be, but also the more data and computer power it will need to be trained.[15] This need was one of the main reasons most mainstream researchers weren't using deep learning for image recognition. But Hinton and his team had been patient. Notably, the techniques used by other researchers, such as those still clinging to symbolic AI, were ineffective at image recognition.

When Krizhevsky started his project, the problem of data had been solved. ImageNet had created a massive database of labeled digital images: more than enough to achieve his desired result. As for computing power, well ... the situation was complicated.

Although CPUs had come a long way, they were not very efficient for training deep learning networks. However, the development of GPUs presented a promising alternative that other researchers had already taken advantage of.

Much like the image-processing tasks GPUs had been designed for, training a neural network such as the one used for AlexNet could be broken down and performed in a series of smaller calculations. This meant that using GPUs to power the system would be much more efficient and practical than using CPUs. By using GPUs, the team would be able to access the required computing power to train and operate their system.[16]

Like I said, though, there was a problem. In 2012, there were no GPU systems powerful enough to train such a large deep learning system. There just wasn't enough demand at the time to make GPUs that powerful.

The team knew that if they could get their system to work, it could justify more research and funding into the creation of more and better GPUs. So, they compromised and worked with what they could find.

Looking back at AlexNet, it is a bit of a jerry-rigged system. The team was working on the edge of what was possible. They knew they could achieve their goal, and the technology was so close, but not quite there. They had to change the system's architecture to work with what they had. Instead of training the whole system on one GPU, they had to find a way to split the system in half and run it on two separate GPUs, which were then sewn back together at the end.[17] It wasn't ideal, but the team prayed that their compromise wouldn't affect the end result too much.

A system like AlexNet takes time to train. It runs through the data given to it, constantly adjusting its nodes until a final model is produced. At this point, either because it has run out of training images or because it has stopped improving, the system freezes and a complex algorithm is produced. Then you can give it new images that it didn't see during training and judge the results.

Up until this point, image recognition algorithms had been crude and ineffective. No one believed that the disciples of the neural network would fare any better. But as the pieces moved, an opening began to emerge. It wasn't guaranteed, but it was a chance. And when the time was right, the queen unexpectedly blasted across the chessboard, taking an unsuspecting pawn. The king was pinned. Checkmate.

When AlexNet was entered into the 2012 ImageNet competition, it did not just win; it demolished its competition.[18] Overnight, all other forms of AI became obsolete. The world now saw what Hinton had always known. Neural networks were the future of computing.

I often wonder what it must have been like when Alex Krizhevsky, Ilya Sutskever, and Geoffrey Hinton first saw the results of their creation.

Did they pop champagne? Did they realize they were destined to become three of the most famous computer scientists in the world?

Did any of them stop and think, *"Dear God, what have we begun? What have we brought into this world?"*

***

After AlexNet, Hinton became a celebrity. He won the Turing Award – the computer scientist equivalent of the Nobel Prize – and was hired as one of the lead scientists for Google's AI program.[19] In 2024, he won the actual Nobel Prize in physics. Yet despite these accolades, Hinton left Google in 2023, expressing regret and fear over what he had helped to create.

In a 2023 interview, he solemnly stated: "*I console myself with the normal excuse: if I hadn't done it, somebody else would have.*"[20]

ACT 3

Frankenstein's Monster

Fear and Hope

CHAPTER 13

Understanding the Monster

Electricity was in the air, both in the form of lightning and the palpable angst that consumed the village people. Rumors had spread through the town that the madman who lived in the castle on the hill was conducting experiments of the most unholy nature. Something had to be done. He needed to be stopped. They had to do something before his meddling with nature doomed them all.

"We must storm the castle," one man declared, slamming his empty mug on the tavern table.

"Burn it to the ground," cried the woman behind the bar, receiving supportive cheers from the patrons who had packed the establishment.

There was no more discussion to be had. The people had seen enough. It was time to act. The hinges of the tavern doors squealed as the horde passionately stormed out. In the hours that followed, the mob steadily grew. Neighbors informed each other of the task at hand. Farmers distributed pitchforks and torches. The moment of reckoning had arrived.

The storm intensified as the mob marched up the hill to the castle. Water streamed down the path, creating great pits of mud, yet the horde marched on. They would not be deterred.

Inside the castle, Dr. Frankenstein finished the final preparations for his experiment. "Make sure the probes are properly placed, Igor!" he barked at the disfigured hunchback who served as his assistant.

Igor, following his master's instructions, methodically inserted probes through a large metal sarcophagus.

"Are the probes in place?" called the doctor, who was standing behind a control panel at the base of the makeshift lab. "We must hurry. We haven't much time!"

"Yes, master," the hunchback replied, his voice drowned out by the symphony of thunder and bloodthirsty howls from the angry mob that grew at the gate.

"Hoist the body!"

"Yes, master."

As Igor cranked a lever, the metal sarcophagus jerked into the air. As it reached the apex of the castle's laboratory, the doctor flipped a switch, and two metal panels fell away from the ceiling, exposing the lab to the elements of the storm.

Wind and rain flooded the laboratory. The doctor and his loyal assistant stood resolute as their precious equipment was destroyed by the water. They did not care. Their mission was too important, and they had come too far.

Bang! The sound of twisting metal echoed through the chamber. The castle gates had given way.

"Barricade the door, Igor!" the doctor cried. "I shall continue the experiment!"

But as Igor ran to the door, there was a second, even louder bang. It was not the sound of twisted steel but instead that of metal and fire. Bang! Igor heard the sound again and turned around just in time to catch the most marvelous and sinister smile cresting the doctor's face as a second lightning bolt struck the sarcophagus.

In an instant, every instrument in the room lit up with a heavenly glow. Electricity pulsed and cracked through dangling wires, jumping from machine to machine. The sarcophagus shook vigorously, and a monotone moan began to protrude from its interior.

As the moan grew into a scream, the doctor cried, "It's alive, it's alive!"

With the press of a button, the wires that held the metal capsule released, and the sarcophagus plunged to the ground. Igor and the doctor raced toward it, crowbars in hand, to free their creation. As they broke the bolts that kept the capsule together, there was one final bang. The door to the lab had been breached, and the villagers swarmed in.

The mob gazed upon their prey. Few of them had ever seen the doctor before. He looked sickly, ravaged by his own obsession. The desire to create something evil had consumed him and left him looking only faintly human. Rain continued to fall into the room, drenching its inhabitants. The mob had frozen in place. They were like a dog chasing a carriage, unsure what to do once they had caught it.

The silence needed to be broken. Action had to be taken. Finding his courage, the man who had first proposed a storming of the castle stepped forward, but as he opened his mouth to speak, an unearthly sound filled the room. A baritone yell blasted from the metal sarcophagus that lay on the floor. Two gigantic green hands gripped the frame, bending the metal as they lifted a horrifying figure into view.

As it stood, it towered over the room's other occupants. Its skin, a lime green, was covered in stitches and scars. With hair of jet black and two metal bolts protruding from its neck, it was the embodiment of all that was evil in this world.

With disbelief in their eyes, the village people hastily began to retreat. Maybe if they ran now, they could still save their families. As the doctor grinned, the imposing green figure leaped from his enclosure and lunged toward the cowering village people.

He had done it. Dr. Frankenstein had made his monster.

***

This is the story of Frankenstein's monster, as it is often told now. He has been relegated and reimagined as a one-dimensional villain. It is not uncommon to think that the monster himself is Frankenstein, because in popular media, there is often little mention of his creator. The monster has become a symbol that we know to fear. But as you may recall, this monster is not the character that Mary Shelley first imagined.

The story of a mad scientist making a monster is a morally simplistic tale. We know who the bad guys are. And we know what we should fear. The monster is evil because it was in its nature to be evil. The villagers tried to stop its creation because they could see that what was being done was wrong. And in many versions, when the villagers fail, a noble hero must slay the monster, freeing the world from its terror.

In Shelley's story, the monster and the world he inhabits are complex. It is not boiled down to a simple story of good versus evil. The monster is not naturally evil and wants to be human. He wants to feel the pleasure of love but is rejected by his creator. It is a story that forces the reader to wrestle with questions of their own ambition and morality. Moreover, it explores how society shapes the things we create. The book is a warning about the consequences of our actions and a reminder that the advancements of science do not occur in a vacuum. They are not neutral. We must think about the effects of our creations before we conjure them, or we may become haunted by their ramifications.

When the problems of AI are discussed, I believe that we tend to talk about them in the image of the simplified story of Frankenstein. There is minimal nuance, and the concept of AI is bundled into a single idea that some think should be feared or controlled. We talk about AI being an existential risk to society, as if the concept of AI itself is the technology we are using.

Don't get me wrong; I acknowledge that AI carries certain risks. In fact, this final section is fully focused on the risks. Nevertheless, I feel that, in the media and popular conversation, we concentrate on the more sensational risks, the ones whose actual impact or likelihood remains uncertain.

When I set out to write this book, my goal was not to write bold predictions about the future. I do not desire to speculate on topics such as the "AI singularity" or "general intelligence" AI systems. It is my goal to share with you the complex ways in which AI is affecting our society right now.

In my line of work, there is constant frustration directed toward the media and government attention to AI's "existential risk" and the need to control it. Like the story of Frankenstein, it preys on our fears of entities that appear to be beyond the normal order of things. To many of my colleagues, these conversations are overshadowing massive issues that are occurring now. These are issues that overwhelmingly affect the most vulnerable members of our societies. They are issues deserving our attention. In fact, there is even compelling evidence that the dialogue surrounding the "existential risk" is in part inspired by industry groups that want to distract from the more nuanced issues of AI. If government resources are tied up trying to deal with the potential future threats of AI, they will not be available to address the potentially problematic foundations of an AI society being built right now.

I don't mean to say that the numerous AI experts who have spoken out saying they have concerns regarding the future of AI, such as Geoffrey Hinton, are wrong. I do not believe that is true. However, I do believe that their voices and concerns have been amplified, and in doing so, we are missing the chance to discuss meaningful issues.

What I have attempted to do in the following chapters is provide context for the issues that I believe are unacknowledged – issues of a changing society, triggered by the launch of AlexNet and the events and decisions that followed. They are issues that you are likely to have already observed in your life but may not have been able to completely understand. What I want to do is give you the tools to understand that the way we have discussed the technology up to now, as an abstract power being developed by outcast scientists, is meaningfully affecting and changing society. What you will read is not original thinking by me. It is not wild speculation. Instead, I see this text as a vessel for the ideas and work of countless academics I will be introducing to you.

Finally, before we dive in, I must make a confession. In the study of technology, it is widely accepted that there is no such thing as non-political technology. All technology is embedded with politics. It is never neutral. The idea of a technology, in the abstract, may be neutral, but in building it, we embed the political into it. Each choice in its design is made for a reason. These reasons have ramifications. The story of AI is a story of design decisions. It is not neutral.

So, my confession is this: I lied to you in the prologue. I told you that I was not going to be political. But the truth is, this book is about technology. Language is an invention of the human mind, and the way we use it is political. It is not my intention to push a political belief on you. However, I have chosen which stories to tell. I decided which facts were included and what was smoothed over or simplified. In writing this book, my values are embedded within it. In many ways, the act of writing a book intended to educate the general public on how AI works is itself a political act, even if I am not telling you how to act. As such, the following chapters cannot be disentwined from my experiences and my beliefs.

So, I think it is only fair that I tell you now where I stand.

I believe that the general technique of deep learning has the potential to change society in amazingly positive ways. But through my research into how it is being integrated into society, I believe that it is currently only amplifying the problems that already exist. I am writing this book because I hope that others might see these problems too and collectively cry, "This is not the future we want!" My political act is picking the problems to highlight to you so that you can see yourself within them. And if you desire, you might act upon them.

To see these problems, we must begin with the aftermath of AlexNet and the blisteringly fast development of contemporary AI systems.

CHAPTER 14

Aftermath

The months and years that followed the launch of AlexNet are best described by one word: pandemonium. Almost overnight, the tech community realized they may have been overlooking the disciples of neural networks. Maybe they were right; maybe all they had really needed was more data and stronger computers.

Remember the feeling you had when you first saw ChatGPT in action. Or the first time you saw an AI program do something truly mesmerizing. How did it make you feel? Did it make you feel scared? How long did it take you to say, "This changes everything"?

AlexNet was the ChatGPT moment for tech companies. It was the moment that thousands of people in the tech world simultaneously got a sneak peek into the future. Suddenly, hundreds of futuristic projects that had been put on the backburner seemed possible. Visions of worlds only before imagined as science fiction danced in the dream of every tech CEO. And the venture capitalists? Well, they smelled money.[1] There was massive untapped potential in deep learning. However, there were also some major problems.

Until AlexNet, the science of AI had been completely pushed aside by the scientific community.[2] Yes, machine learning (where humans provided the parameters) was a valuable tool, but it was associated more with statistics and math than the field of AI. Moreover, within the field of AI, neural networks were considered a fool's game.[3]

Although this book has focused on Hinton and his team, they were not the only ones who remained dedicated to neural networks. Around the globe, there were small pockets of labs and researchers who remained convinced that neural networks would eventually work, and although they were not part of the AlexNet team, AlexNet would not have been possible without their research.

So now, there is a huge desire for more deep learning but not many people who are actually experts in it.

What followed was a bare-knuckle street fight between major companies, battling to acquire the limited human resources that existed in the field.

There were a handful of big tech companies that dominated this fight. They are the usual suspects – Google, Facebook, Amazon, and so on. These companies had four major advantages: (1) money, (2) data, (3) computing power, and (4) history.

The variable of money is a pretty straightforward one. With only a limited number of deep learning experts in the world, and intense competition to hire them, the price of wooing one was quite high. Companies competing with one another pushed the price up, making it difficult for smaller tech companies or universities to compete. It wasn't just salary. Once you brought these researchers in, you needed to provide them with the funding and staff to support their visionary work. Although researchers like big salaries, they are often more attracted to venues that will pay the big bucks to support their research and make it a reality. Even if deep learning tech wasn't fully ready to be rolled out to the public, the big tech companies could afford to eat short-term losses – with the promise of future innovation. But money couldn't easily buy everything these researchers needed. Fortunately, these companies had a natural advantage.

Training deep learning systems takes a lot of data. And data is created through the act of surveillance. Companies such as Google, Facebook, and Amazon were already surveillance-based businesses. Because of that, they had a LOT of data, which gave them a massive competitive advantage. Furthermore, due to the nature of their businesses, they also already had the powerful computing infrastructure necessary to support these projects.

The final advantage they possessed was that of history. Google specifically has always seen itself as an AI company.[4] Their current CEO has explicitly stated as much, saying in a 2023 shareholders meeting, "AI has been foundational to our Ads business for over a decade."[5] Although they were not using deep learning algorithms, their business model relied heavily on processing massive amounts of information to achieve a goal. Whether that was the PageRank algorithm or the real-time auctions, they had long tried to create systems that could perceive the environment around them and use the information they gathered to achieve their given objective.[6]

Before AlexNet, Google already had numerous AI projects.[7] They even had a small research team working on deep learning technology.[8] Waymo, their self-driving car project, began in 2009, three years before Hinton's team demonstrated the power of deep learning for image recognition.[9]

It is also important to note that, by this point, Google looked nothing like the search company from chapter 10. They had drastically expanded their mission and were one of the largest companies in the world. In 2012, they were internally researching and developing robotics and biotech technology.[10] They had also already publicly joined the consumer electronics market, releasing Android, the most popular smartphone operating system in the world, in 2008.[11] They released their first phone in 2010.[12] And they acquired Motorola, a consumer technology giant.[13] But their expansion did not stop there. In the years directly following AlexNet, they would make major moves, including purchasing Nest, a smart home technology company.[14] They also started a pharmaceuticals company in 2013,[15] and doubled down on their health sciences division, spinning it off into its own subsection of the company called Verily in 2015.[16] Needless to say, Google has big ambitions. Yet their most important acquisitions for our story occurred in 2013 and 2014.

Following the success of AlexNet, Krizhevsky, Sutskever, and Hinton started a new company called DNNResearch Inc.[17] But the company didn't last long as an independent startup. Not even a year after its founding, Google quietly acquired the company. As part of the acquisition, Krizhevsky and Sutskever moved to Silicon Valley, and Hinton agreed to split his time between working at Google and continuing his research at the University of Toronto.[18]

The group provided immediate benefits for Google, helping to launch a smart photo feature that leveraged their image recognition AI. But the real value they provided was long-term. Their knowledge and expertise aided Google in staying at the cutting edge of deep learning.

A year after Google acquired DNNResearch Inc., they acquired another major AI company, DeepMind.[19] Although not as flashy as AlexNet, DeepMind had developed some of the most impressive deep learning systems of the time.[20] Acquiring the company and their talent pool paved the way for Google's future innovation in the AI space.

These were not the only two companies acquired by Google during this period, as they periodically scooped up dozens of small startups. Of course, they were not alone, as the other major players followed suit.

In the period between 2012 and 2022, AI research and development silently barreled forward right under the public's noses, yet we never took notice. Hundreds of new deep learning techniques were pioneered that allowed for faster and better neural network training. The paper trail is clear, and companies did not try to hide it. The fact is, we didn't care, because we could not yet tangibly visualize the power of the transition that was playing out.

Looking back at this time period now, it seems odd how little we talked about the rise of AI as it happened. I was in Mountain View, California (where Google HQ is), during the summer of 2016, and I couldn't believe my eyes when I saw a self-driving car on the road. I still have the picture of it on my phone. During the week I was there, the sight of the self-driving cars became mundane. I was blown away by the technology, but I never stopped for a second to think about the greater implications.

But it wasn't just prototype cars; AI became embedded in our everyday lives, and we didn't even notice. In 2015, Snapchat launched their lens feature.[21] With a click of a button, the user could transform their face, shooting rainbows out of their mouth or adding adorable puppy dog ears. That was AI.[22] They used deep learning to be able to recognize facial features in real time and alter them.

A common complaint when I was in my undergrad was that the Marvel movies were starting to feel formulaic. In fact, all of Hollywood began to feel cold and calculated. Yeah … it was AI. It became common practice to run big budget movie scripts, especially Marvel movies, through AI programs that claimed to be able to predict the box office success of the film. The AI system would then suggest changes that could result in bigger box office numbers.[23] "Cast Brad Pitt in this role and increase earning potential by 7 percent." "Add another fight scene in the second act and up earning potential by 10 percent." The company that operates this system tries to keep a low profile. But if you are curious, its name is Epagogix.[24]

Google ads seem to be getting scarily accurate? AI. YouTube and Instagram feeds became more tailored and more addictive? AI.[25] News recommendations become more tailored to your political beliefs? AI. Suddenly we can interact with our devices by talking to them?! AI. Like image recognition, deep learning techniques provided an avenue for voice to text to become much more practical.

These are just some examples of AI systems that you may have interacted with, without realizing. The fields of medical science, and especially finance, became overrun with AI applications.[26] But we didn't talk about it. We didn't notice it because the technology seamlessly integrated with pre-existing systems. Voice recognition existed before deep learning. It just suddenly got better. We were already getting tailored ads. They just got creepier. Epagogix has been making movie algorithms since 2003. They just got more all-encompassing with the addition of neural networks.[27] In fact, I did a project in grade 7 on self-driving cars. That is part of why I was so excited to see one. They weren't a new idea – they had just suddenly become more practical.

To the public's and media's eyes, computers were just doing what computers already did, only better. But at some point, we started making programs that could do something that we didn't expect computers to do – something that felt very human. We taught them to write.

CHAPTER 15

GPT

When I first learned about the GPT project in 2019, I was intrigued. A program that could write anything, just from a prompt, and could easily pass as human-generated content? It seemed too good to be true. But, as more information seeped out of the project, I got to see more examples of what it was writing, and it looked, dare I say, good. Eventually, I got my hands on a demo that could write 250 words from a prompt. The results were promising but at times very underwhelming.

My research at the time was focused on the potential political ramifications of AI systems that could write like humans, but admittedly, I became hesitant about the timeline for the effects I was studying.

"It's going to take years before this really starts affecting society," I thought to myself. But my work was starting to garner attention, so I stayed the course, while also broadening my knowledge on AI in general. Then, when preparing for a presentation on text-generation AI systems, two events occurred in blistering succession. Suddenly, I realized I was wrong; we really didn't have much time left.

First, I read a *New York Times* article about Bloomberg News. At the time, text generation still seemed like something that existed only in research labs. But according to this article, one-third of all articles published by Bloomberg were being written by robot reporters.[1] This was 2019. But it wasn't only Bloomberg; outlets such as *The Guardian* were following suit.* I had probably read countless AI-generated articles and never even noticed.

Second, OpenAI announced GPT-3.[2] Not ChatGPT. GPT-3 was the successor to GPT-2.[3] The GPT project was the company's venture into creating a natural language processor, a deep learning AI system that can comprehend and navigate human language. One of the major outputs of these systems is the ability to produce text from a prompt. It is helpful to think of the GPT systems as an engine and the projects built around them as different cars. The first publicly launched ChatGPT system ran on GPT-3.[4]

From the moment OpenAI released their initial reports on their GPT-3 project, it was obvious that it was a completely different beast from GPT-2.[5] It was the difference between the engine of a Ford Model T and a modern Ferrari. When I finally got my hands on a demo that demonstrated its text-generation capabilities, it was clear there was no going back. Once this system was polished and public, everything was going to change.

But let's take a step back. Where did all this come from? Up until now, the name OpenAI has been completely absent from our story. Although that is true, the characters who made it possible were not. To understand the rise of OpenAI, we need to go back to where it all began: AlexNet.

ILYA AND GENERAL INTELLIGENCE

Ilya Sutskever was one of the three creators of AlexNet and a cofounder of DNNResearch.[6] When DNN was acquired by Google,

---

* Although it now takes a strong stance against using AI in journalism, *The Guardian* was an early adaptor of the practice, experimenting with the program ReporterMate.

Sutskever moved to Mountain View and joined Google Brain, one of their AI divisions.[7] While there, he continued to demonstrate his potential as an AI scientist.

Google entrusted him to work on numerous open-ended AI research projects. He played a vital role in the development of the tools needed to build deep learning systems. It was during his time at Google that he started developing the foundations of a new idea. Maybe the same pattern recognition functions that worked so well for pictures could also be used for words. Unfortunately for Google, his tenure with them would not last long.

In 2015, Sutskever was convinced by Elon Musk to leave Google and become a founding member of a new "non-profit" AI research center: OpenAI.[8] Sutskever was to be the chief scientist.

I put "non-profit" in quotations because, although that was the original mission, it is no longer the case.[9]

OpenAI's original mission was to build "safe and beneficial artificial general intelligence for the benefit of humanity."[10] Up until now, I have avoided using the term "artificial general intelligence" to avoid confusion. Alas, it now seems inevitable.

I have avoided using the term because it at once means a lot and absolutely nothing.

There are just too many different definitions for artificial general intelligence (AGI). The concept has its roots in the early computing work of Alan Turing. If you recall, it was Turing's goal to build a general computer. A general computer would be a computing device that could perform a wide range of computing tasks, not just one narrow function. His codebreaking machine was not a general computer because it could only perform one task. Babbage's Analytical Engine, on the other hand, could be reprogrammed to do multiple functions, so it was a general computer.

The concept of AGI refers, in the abstract, to an AI system that can perform a broad range of tasks, not just a narrow task like predicting if a movie will make money.

The problem is, at what point do we actually deem a system to be generally intelligent? How do we measure intelligence? Turing provided one possible solution to this problem through the Turing Test, proposing that, since intelligence is a human concept, we should measure it by the ability to mimic a human. Although the Turing Test is popular in science fiction, it really only measures a specific form of human intelligence.

In fact, although I ignored this point in chapters 6 and 7, AGI was the goal of the early AI pioneers. They believed that the way to reach AGI was to break down the abilities that made humans intelligent, and tackle them individually, before eventually combining all the systems. Thus, they built logic systems, image recognition systems, and so on.

Here we also run into a serious problem about AI that you are sure to encounter as you explore this world. The definitions are a hot mess. It makes having intelligent conversations very difficult. I ran into this problem a lot during my early days researching.

For a lot of researchers, the term "AI" refers to general AI. Systems such as deep learning are just very advanced machine learning systems. To be an AI, you need to be general. But even that definition is hard to reach because the target for what is general AI is a moving target. For some, it's passing the Turing Test; for others, that is only a start.

Other definitions place AI under a broad umbrella, with lots of different techniques such as machine learning and symbolic approaches under it. Under this system (which I use in my work), AI is a much broader concept. Ideas such as general AI, or artificial consciousness, are goals that may be achieved through the development and construction of AI systems.

I will also be honest and blunt: I dislike the concept of AGI. As I stated earlier, I don't like systems that center human intelligence. I especially don't think we should use human intelligence as a measuring stick for the success of a computer. Because a computer is … well … not human.

The cuttlefish – one of nature's most fascinating creatures – can detect the color and shape of its environment and change its body shape and skin pigment to camouflage itself. Some of them are better at camouflage than others. That is crazy smart. Can you do that? We should not discount the intelligence of our dear friend the cuttlefish just because its structural attributes are different from those of humans, lending itself to different types of intelligence.

Computers are not humans.† They are built differently, and we shouldn't judge their intelligence as if they are. In fact, there are already so many tasks where AI-powered computers are far more capable than humans due to their structural qualities – they possess their own intelligence strength, much like our dear friend the cuttlefish.

Anyway … back to OpenAI. OpenAI's goal is to develop general intelligence; fortunately, they do provide a definition. To OpenAI, AGI represents "highly autonomous systems that outperform humans at most economically valuable work."[11] The bulk of OpenAI's research is directed at deep learning, and they have been at the forefront of developing groundbreaking AI systems and training architectures. But of course, what they are most known for is ChatGPT.

Before we dive into ChatGPT, I will take a quick detour to present an AI training technique I first learned from OpenAI. I want to acknowledge it, even though it does not cleanly fit into our narrative, as I think it is important for you to understand how training techniques have evolved beyond the models used by Hinton and his team.

The technique of competitive self-play has become an extremely popular model for training AI. In the deep learning systems we

---

† Duh!

have explored up until now, the system is provided with the data to train from. However, in a competitive self-play system, the AI produces its own data.

The first competitive self-play model I saw was produced by OpenAI and was honestly quite amusing.[12] They created a virtual sumo wrestling game. In the game, two stick men stood on top of a virtual platform. Each stickman was controlled by its own AI system. Their goal was to win the game by pushing the other stickman off the platform. There were no other instructions. When one stickman fell off the platform, the survivor was given a digital reward to tell them they had done a good job. Then, the game would be reset, and the two virtual stick men would fight again. This scenario was repeated hundreds of thousands of times.

During the first matches, the winner was basically random. Whichever stickman accidentally stayed on the platform won. But over time, one of the stickmen realized that, when they accidentally bumped into their opponent, knocking them off the platform, they were rewarded. This reward strengthened the connections between the nodes in the neural network that directed the stickman to charge at their opponent. The stickman had learned a strategy: hitting. Slowly, over thousands of simulations, new strategies emerged. The stickmen learned to bend their knees, and by the end of the simulations, they even developed complex wrestling techniques. The miraculous part: even though we can now go back and watch all of these matches individually, the real-time simulation of all the matches took mere minutes. Through competitive self-play, AIs can train themselves by competing against other AIs. We call models like this "unsupervised learning," since humans are not helping the AI understand the data.

The most famous example of competitive self-play is Google's AlphaZero chess engine.[13] Prior to AlphaZero, chess was often pointed to as a great example of why AI would not take our jobs. Although chess algorithms had developed to the point where they could consistently

beat top chess players, when chess players teamed up with an AI algorithm, they produced an even more formidable team. Yes, AI was
scary, but it was just a tool, and we were better off using it to improve
our existing abilities than completely automating ourselves.

But before AlphaZero, chess engines had been trained on a
dataset of past chess matches.[14] With AlphaZero, the engine was
provided with the rules for chess and was instructed to play 44 million games against itself.[15] It took nine hours to train.[16]

AlphaZero has never been beaten.

More worryingly, when humans play on a team alongside
AlphaZero, it actually diminishes its effectiveness. AlphaZero
is better on its own. It learned chess organically, unaffected by
human assumptions about what the best way to play was. It makes
opening moves that any human chess player would be laughed at
for attempting and turns them into ruthless strategies.

AlphaZero created new chess strategies and moves: these were
moves that no human had ever thought of.[17] Here we witness the
death of Lady Lovelace's Objection. Recall how she stated that
a machine could never create something new, a claim Turing
refuted. Well, I cannot think of a clearer example of her misjudg-
ment than this one. Whether we like it or not, the machines can
now create. They are no longer rearranging ideas that we have
already had, but instead, through the statistical architecture of
neural networks and techniques such as self-play, they can genu-
inely make discoveries. This is a paralyzing realization for some.

Around the same time as the AlphaZero work, Google AI researchers, in collaboration with researchers at the University of Toronto,
made a groundbreaking AI discovery.‡ In June 2017 (a year and a
half before AlphaZero), they published a paper titled "Attention Is

---

‡ It is refreshing to see that the University of Toronto seems to have replaced Cambridge as the spoke in the wheel of AI development. As a Canadian, that makes me
   unreasonably happy.

All You Need."[18] In a ranking of the most important scientific papers ever written, "Attention Is All You Need" finds itself in the company of Turing's early work. In the paper, the researchers proposed a new kind of neural network architecture called a transformer model.

The transformer model was designed to address the difficulty that traditional neural network models were having with text-based tasks. The paper focused specifically on the task of translation. Before the transformer model, the primary architecture used for text-based tasks was called a recurrent neural network (RNN) model.[19] RNNs looked at each word in a text one by one, which caused a lot of problems with understanding the context around the words. Also, RNNs couldn't remember much of what they had translated before, leading to even more problems with understanding the context. I remember using early text-generating AI systems (predecessors to ChatGPT) that were powered by RNNs. Because they lacked the ability to remember previous chunks of text, they often rambled and got off topic extremely fast. If I asked one of these systems to write a story about George Washington, it might only have been able to write two or three coherent sentences before changing topics to the politics of Washington State.

The translations RNNs made worked okay, but they didn't capture the subtle details that human translators can.[20] Plus, it was really hard to make RNNs learn faster by breaking their training into smaller parts that could be done at the same time. Because you couldn't speed up their training this way, teaching RNNs was a slow and expensive process.

The transformer model introduced two new big ideas. The first is called positional encoding. Imagine tagging each word in a text with a little note that tells you where it stands among other words. It helps the computer to understand patterns in how the words are lined up. The second big idea is the attention model. While positional encoding helps during learning, the attention model helps the AI pay more attention to certain words when it's working on tasks like translating. For example, in translating, this model helps the computer consider how nearby words affect the meaning of the word it's translating. It makes the translations much better because it helps the computer understand how words are arranged differently in various languages. The computer learns which words are important by practicing on texts that have already been translated. This practice uses something called self-attention, where the computer figures out by itself which words to focus on as it learns. Lastly, the transformer model is better than the old ways because it can be trained faster and cheaper, which means we can use much bigger amounts of data to teach it, making it even better.

The paper showed very promising results when applied to the task of translation. But when OpenAI saw the paper, they thought even bigger. Up until this point, natural language processing (NLP) AIs were built for a specific purpose, such as translation.[21] This is in part because, much like the image recognition systems we previously explored, the data needed to be labeled for the machine to learn from it. A human needed to go through the data and annotate it, a time-consuming and expensive task, which greatly limited the amount of data available for use. Therefore, programs needed to focus on specific tasks so that a narrower amount of data was needed for training.

But the researchers at OpenAI theorized that they could potentially leverage the transformer model to change this limitation.[22] They could make a general system, trained on the vast amount of unlabeled text data that was available, develop an understanding of language, and then fine-tune the model for specific tasks.[23] The result might be a central engine of sorts, which lots of different types of vehicles could be built around, a pre-trained model that could be built on to perform numerous text-based tasks – a generative pre-trained transformer, or GPT for short.

The first GPT they made was a proof of concept.[24] They took the architecture from the transformer model and trained it on a massive amount of unstructured data. Although it wasn't perfect, it was a great proof of concept.

The second GPT, the one I first saw, built on the proof of concept and attempted to make it more practical.[25] They trained it on a much larger data set and increased the number of parameters. More parameters basically means a greater capacity for memory. If we imagine the layers of a neural network making the system deeper, more parameters make it wider, giving it the potential to infer more patterns and create more weights between each layer. The more parameters there are, the more nuance the model can develop.

GPT-2 was a smashing success.[26] It began to signal that something big was on the horizon. GPT-2 had 1.5 billion parameters, making it an unthinkably powerful machine.[27] For their next act, OpenAI would up the ante an unprecedented amount. Not only was GPT-3 to receive more training data, but it would have 175 billion parameters.[28]

If you recall, it was my introduction to GPT-3 in 2020 that really scared me into seeing how fast AI was developing. Although you may think that your first encounter with GPT was ChatGPT, that is probably not the case. Following a period of delays over safety concerns, which I am on record as supporting, OpenAI made the decision to release GPT-3 to the world.[29] They also launched what is known as an application programming interface (API),[30] a program that allows two applications to integrate with one another. Now, for a fee, other applications could harness the power of GPT-3.[31] For three years leading up to the launch of ChatGPT, you probably unknowingly interacted with GPT systems hundreds of times.

In 2022, OpenAI released an updated version of GPT-3 called GPT-3.5.[32] GPT-3.5 had updated training data and some fine-tuning that made the system noticeably better.

An interesting by-product of the GPT project was that it seemed to be able to do things that it hadn't been trained to do. It was uncomfortably good at translating languages, even though it hadn't been trained for translation. Moreover, with only a little tweaking, some early users found ways to make it write code based on a text prompt.[33] The researchers at OpenAI eventually tried a new idea, taking the model and seeing if it could recreate images. The project was a shocking success and was named DALL-E.[34]

I often showed off my GPT-3 and GPT-3.5 access as a nerdy party trick. Honestly, I was often disappointed that no one found it as shocking as me. But that wall started to crack when systems similar to DALL-E became widely available. In 2021, a machine learning consultant who was inspired by OpenAI released a free web application called DALL-E Mini.[35] The website was very easy to use and could generate a set of nine images from any prompt. It ignited the imagination of millions and quickly became the go-to generator for internet memes.

This was the first time my friends, avid meme lovers, really cared about AI. They spent hours creating every funny situation their brains could possibly think of: Darth Vader Zerg, security camera footage of Shrek robbing a convenience store, a pug storming the beaches of Normandy – it was the pandemic, and we were very bored.

But after a few months, the hype wore off again, until November 2022, when OpenAI dropped an atom bomb: ChatGPT.[36]

I do not need to tell you what ChatGPT is. You have almost definitely used it. You type a prompt, and the machine answers you. ChatGPT changed everything for AI. Not because it was revolutionary; we had had the basic technology for a while. But because it was unmistakably AI and terrifyingly powerful: suddenly no one could ignore AI.

I was working as a teaching assistant marking papers at a university at the time. Before ChatGPT, a lot of the academy didn't take my work seriously. "You are wasting your potential," I would hear. "Work on something that exists and isn't just a science fiction pipe dream." "AI will never be able to write papers." But then one day I woke up – and all we could talk about was AI. Teachers were terrified that students were using ChatGPT to write their papers. As the one who marked those papers, I can guarantee they were, because a lot of those kids got much better at writing after that November.

The next six months were wild. All the media could talk about was AI. Every public policy forum was on AI. Joe Biden was talking about AI. Many brushed it off as hype, saying ChatGPT was a parlor trick. "Look at how, right after ChatGPT, everyone is suddenly doing AI! It's obviously a scam!" I heard the critics shout and tweet. "It will die down in a few weeks."

But it didn't. ChatGPT got even better with the release of GPT-4.[37] Image-generating AI got scarily good, and every day the news reports on a new AI innovation.

I get why people thought it was just hype or a scam. It seemed to come out of nowhere. And don't get me wrong, there is a LOT of AI hype. And there are a lot of new AI companies capitalizing on it that are not going to work out. But it didn't come out of nowhere, and it didn't start with OpenAI or ChatGPT.

For eleven years, from the launch of AlexNet to the launch of ChatGPT, billions of dollars had been invested into AI. Google, Facebook, Microsoft, Apple, and Amazon all invested billions. The AI products you are seeing now are not rushed. You have been interacting with AI for way longer than you were aware. But before ChatGPT, no one cared. So companies didn't really advertise the fact that they were using it. It didn't seem to matter.

But now, the genie is out of the bottle, and it's good marketing to let everyone know exactly how much AI you are using. Honestly, right now, it is a little over the top, and companies do often lie about using AI. But as the hype settles down, the technology will remain.

We now live in an AI society, and that is changing things. There is no sugar-coating this fact. Things are changing. Not in the future. Not in the near future. Right now.

CHAPTER 16

Beyond Generative AI

AI is not one single thing. It is not ChatGPT, and it is not just deep learning systems. It is a series of technologies that interact with each other and other non-AI tech to achieve goals assigned by humans. A lot of the time, when people talk about AI, they are referring to generative AI, made from large language models like GPT-4. That is the public perception of AI.

I understand this perception completely; generative AI is flashy, and it is very visible. But it is not the only type of AI in action. As we transition into discussing the effects of AI on society, I feel it is vital to address this point and hammer it home.

In this book, we have already discussed countless other forms of AI. Before deep learning, there were systems that helped humans make decisions by applying expert knowledge and logic skills. These systems didn't need to create anything new in the way that generative AI tries to do. Their job was to apply existing information to a situation.

We may also look at image or voice recognition. Both use deep learning but aren't generating an organic output. The AI algorithms used by YouTube or TikTok to curate your feed are using deep learning to carry out predictive tasks. Their job is to try to make an accurate prediction based on the available data.[1]

AI is being used in so many different places now. In health care, it is being used to detect cancer and to read scans.[2] It is being used by the IRS to try and detect white collar crime.[3] It is being used to predict and model weather,[4] schedule surgeries,[5] and improve farming production.[6] It is also being used in more controversial ways. Facial recognition is being used to identify and intimidate protesters.[7] Ukraine is currently a testing ground for autonomous weapons.[8] And it is being used by police forces to try and predict where a crime will occur and stop it before it happens.[9]

However, it is important to remember that the things we call AI are not always just one type of AI, or one piece, or solely AI at all. Computer vision, a major field in AI, focuses on giving computers the ability to comprehend their physical environment, giving the computer eyes.[10] To do that, you need to run deep learning software similar to that used for image recognition. But the algorithms themselves are completely useless without the cameras and sensors they are built to work with.[11] What one may call an AI camera is actually a collection of technologies working in tandem to create the illusion of intelligence.

For some AI systems to work, multiple types of AI and non-AI technologies are necessary. Self-driving cars are AI systems. The car can perceive its environment and make intelligent decisions based on its perceptions. Yet, in order to perceive its environment, it needs its sensors, and to understand the data from its sensors, it needs a computer vision algorithm. But not even that is enough. Once it can understand its environment, it needs to make decisions. These decisions need to take into account the rules of the road, driving norms, and the events occurring around it – another

AI system. None of this would work without the thousands of other pieces of non-AI technology that make up the car.

When we talk about AI systems, it is easy to get caught up in the idea that AI is a monolithic thing. I am guilty of this myself sometimes. But it isn't. That's not how technology works.

AI systems are assemblages of sensors to perceive the environment, algorithms to direct their response, and other technologies that execute the instructions given by the algorithms.

Such systems might be small and self-contained, like a smart assistant device that fits in your pocket. Others can stretch across entire countries or continents. In the next chapter, we will explore one such system, which is gradually changing the way humans work, how we live our lives, and even how we perceive the world around us. A technology that was the desire of tech companies long before the invention of deep learning, that is now becoming possible. It is a vision of the future that has already begun to take hold. This vision of computing is looking to move past the two paradigms of computing that Steve Jobs ushered in to completely reframe how humans and computers interact.

CHAPTER 17

The New Computer

Mark Weiser is often called the East Coast Steve Jobs.[1] The two share a striking resemblance, not by virtue of their looks, but due to a resemblance of minds. They were both visionaries who saw technology not as it was, but as it could be, and who understood that, at the center of every technical system, there is a human who needs to use it.

In his early twenties, Mark Weiser was accepted into the master of computer science program at the University of Michigan. This achievement might not raise eyebrows in a book filled with computer science prodigies – until one discovers he never obtained an undergraduate degree. Like Steve Jobs, Weiser was a liberal arts dropout. He spent three semesters pursuing philosophy before he ran out of money and abandoned his studies.[2] Nonetheless, true visionaries are hard to keep down for long. Weiser swiftly taught himself the skills required to excel in the emerging field of computer science. He started with an entry-level job as a programmer and worked tirelessly in his downtime to improve his skills.[3] Yet, even as he dedicated his professional career to programming computers, his heart remained loyal to philosophy.

Mark Weiser was unique among engineers in that his passion was not for machines but for humanity and our peculiar existence. He was driven by ideas about humanity, focusing on questions not merely about creating supercomputers, but on developing computers that enhance human life. This perspective shifted the dialogue from technology for its own sake to technology as a means to serve human needs more effectively.

At some point, Weiser came across the philosophy of Martin Heidegger, a legendary German philosopher.[4] Heidegger had a complicated history with technology. He did not condemn it as being purely bad, but he certainly did not believe it was good. To Heidegger, technology mediated how humans experienced the world.[5] How we designed our technology would dictate how we understood ourselves and how we behaved. Thus, it was vital that we design technology that would promote living a good life and strengthen our connections to the things that make life worth living. Heidegger challenged the notion that technological progress was progress at all. In fact, to Heidegger, some of humanity's greatest inventions seem to have hurt more than they helped.

The work of Heidegger frightened Weiser. As a professor of computer science in the late 1970s and early 1980s, he had a front row seat to the rise of personal computing. To him, the desktop created by Jobs was an abomination. It didn't free humans. It trapped them at their desks.[6] Weiser fretted that, just as Heidegger predicted, the machine would come to dominate our lives, and we would slowly become less human. He couldn't just sit back and let it happen. Something had to be done. But he wasn't going to change the direction of computing by teaching at a university. He had to get his hands dirty. So, in an attempt to save us all from Steve Jobs, the East Coast Steve Jobs moved west to Silicon Valley.[7]

Weiser went to the Valley to work for a familiar name in our story: Xerox PARC, the same company that developed the mouse and the GUI, only to lose them to Steve Jobs. He quickly rose through the ranks, and after nine years with the company, he became the chief technology officer.[8] His time at the company was a philosophical revolution for computing. He wasted no time dwelling on the company's past success developing technology for desktop computing and instead suggested an entirely new philosophy for the computer itself. A philosophy, of course, inspired by Heidegger. Weiser convinced PARC to ditch the idea of the desktop and completely reimagine what a computer could be.[9]

A philosophical thought that seems to have influenced Weiser's approach to computing was Maurice Merleau-Ponty's musing about a blind man and his cane.[10] Merleau-Ponty was influenced by Heidegger's work and used the cane as a metaphor to explore the nature of technology. The blind man's cane is a fascinating example, as Merleau-Ponty notes that, through its use, the cane becomes an extension of the blind man's body itself. The cane adds to his being in the world, extending his senses. Merleau-Ponty and Weiser were fascinated by the cane because it was not just a piece of technology that a human interacted with, but instead it was a piece of technology that became invisible to the human as it became integrated into their ways of perceiving and interacting with the world. The cane became a transparent extension of the blind man's body.

Weiser thought that computers could do the same.[11] They could become a technology that disappeared seamlessly into the background of human life.[12] Instead of a single desktop that acted on its own, Weiser dreamed of an interconnected ecosystem of tiny computers. The idea of a computer would no longer be tied down to a singular machine. It could be fluid and exist all around the user. The computer should exist in the real world, he thought, not as a distraction from it. He called this computing philosophy ubiquitous computing.[13]

Sadly, Mark Weiser passed away from cancer in 1999.[14] He was forty-six years old. When he died, technology still had not caught up to his mind. The internet was still young, and computers weren't yet cheap enough or small enough. There were some small successes. Xerox Labs created lots of prototype technologies, and the idea of smart spaces gained some popularity. Fortunatelies for Weiser, his ideas were so powerful that they outlived him. His writings – and those he inspired – lived on. They inspired a new generation of innovators, a generation that aspired to make his dream a reality.

The road to ubiquitous computing has not been smooth. In the abstract, it has had some success as, in the early 2000s, the size of computers shrunk again, and the invention of radio-frequency identification (RFID) made the integration of tiny computers into more spaces practical. But it remained more of a novelty and didn't redefine how we interacted with the computer itself.

In the late 2000s and 2010s, the idea got a major boost from the invention and popularity of the smartphone. Suddenly, computing power was constantly at our fingertips. Yet, it still did not live up to Weiser's vision.[15] The smartphone revolution just shrank the desktop computer and made it more accessible to us. It may have become an invisible extension of our bodies, but not in the way he had hoped. Instead of seamlessly blending into our environments, the smartphone has become a distraction from them.

Even though smartphones were not exactly in line with Weiser's vision, our everyday lives were slowly invaded by a lot of small computers. These computers could often talk to one another. Smart devices began to pick up momentum, and the concept of the Internet of Things (IoT) was born.[16] Picture this: a world where almost everything around you, from your toaster to your bicycle, can chat with the internet and each other just like your smartphone does. That's IoT – a massive network of gadgets and everyday items connected online, sharing information to make our lives easier and smarter.

I grew up in a house of tech-obsessed programmers. One of them worked for a major Internet of Things company. We had every smart gadget you can imagine: smart light bulbs, a smart thermostat, smart smoke detectors, and a smart assistant and control pad to run the whole thing. I even recall a smart kettle at some point! I cannot lie – although I am hesitant about smart devices, I do have a smart coffee machine that will make my coffee before my morning alarm goes off. And it is a very important part of my life.

Of course, the concept of smart spaces did not stay isolated inside homes and offices. Governments and municipalities quickly caught on to the potential value of the IoT and began implementing smart city projects. The idea was that, by installing IoT devices across an urban center, municipalities could monitor the city and better use and allocate precious resources.[17] In cities around the globe, energy systems are run by smart grids. Traffic light systems are connected to thousands of tiny monitors and cameras that track the flow of cars. They can supposedly alert authorities to potential issues and help in policing and surveillance of dense urban environments.

In a sense, smart cities were the closest thing we had to Weiser's vision of ubiquitous computing before AI* – an environment where the computer blended in and enhanced human experiences without dominating them. It is possible that you have been to a smart city and never noticed. New York City and Washington, DC, are widely regarded as two of the "smartest" cities in the world. Other smart cities include Barcelona, Tokyo, Singapore, London, and

---

*   That is, when they worked. Although they may sound utopian, smart cities have a tendency to not actually be that smart or really that useful at all.

Amsterdam.[18] In some cities, such as Singapore, the presence of tech is very visible. It is impossible not to notice security cameras on every street corner and robotic police dogs patrolling the streets. But in a city like Barcelona, the computers have been designed to be invisible, deeply embedded into the cityscape.[19] Although the data they produce and manage have a massive impact on everyday citizens' lives, it is possible that citizens are completely unaware that the smart city program even exists.

Although smart cities are a good example of ubiquitous computing at the city level, they didn't fully live up to Weiser's vision. At the personal level, computers were still an object that acted as a distraction. As I sit on a metro in a smart city, although the city may be adapting the metro schedule to better address my commute, I am still probably looking down at my phone and blocking out the world around me. Weiser's ambition was for the computer to truly exist all around us, adapting to our individual and collective needs – for it to only be visible when it actually suited our desires.

Weiser didn't often mention AI in his work. Yet curiously, in one of his most famous writings on the subject of ubiquitous computing, he brings it up.[20] In the set-up for a discussion paper on the ubiquitous technology Xerox PARC was developing, he notes that there are already hundreds of invisible small computers in every house, hidden behind light switches, thermostats, and ovens. Curiously, these computers (at the time) do not know what room of the house they are in. They don't have context on their location or surroundings. Weiser proposed that, by simply informing the computer what room it was in, it could adapt its behavior in significant ways "without even a hint of Artificial Intelligence."[21]

Weiser was attempting to build his computing environment without the aid of AI. This period was, of course, during the late 1980s and early 1990s, a notable AI winter. Another technology he was skeptical of was virtual reality (VR). As he saw it, VR was a grotesque distraction from the real world.[22] Yes, when in the virtual world, you are completely surrounded by computers. But at the same time, you are completely cut off from everything actually around you. It is thus ironic that AI and VR are the driving forces behind a contemporary ubiquitous computing revolution.

During his lifetime, Weiser popularized the concept of ubiquitous computing and witnessed it getting warped by those who misunderstood his vision.[23] He desired a world that reimagined our relationship with computers, where the computer disappeared as much as possible. But instead, his philosophy and inventions were constantly perverted to transform ever more of human life into computing in order to build a more digitized world. In his final years, he began to mount a defense to what he saw as a perversion of his vision. But when he died, his defense died with him. Now his concept has taken on a life of its own. Now it has become a mission to embed humans deeper into the infrastructure of machines. The new vision of ubiquitous computing imagines the destruction of the computer as a single device and reimagines it as an omnipresent experience that flows through and around its user: always present and always adapting. The development of artificial intelligence systems has opened the door for this version of ubiquitous computing to become reality.[24] Sadly, Weiser's legacy may actually be his nightmare.

Reading Weiser's work, I am confident that he would have scoffed at those who insisted that the mobile computing era is a realization of his vision. He was a pioneer of tablet technology but had a fundamentally different idea of how it should be implemented into the computing environment.[25] He spent his final years trying to separate his original vision for tablet computing from the one that would eventually evolve into Steve Jobs's iPhone.[26] Yet, I find myself unsure of how he would feel about contemporary ubiquitous computing, the wave of computing now replacing mobile computing. I think his reaction would contain multitudes – and not be clear cut.

AI is increasingly freeing us from the need for touch screens, keyboards, and mice. Although these technologies are likely to remain in some capacity, just as the keyboard did not die with the touch screen, they are losing relevance. Advances in voice recognition and natural language processing are making it easier to control a computer with one's voice, the same way we communicate with other humans. Although still not perfect, I have found myself increasingly interacting with my devices through voice features over the past year. I have also had the privilege of demoing some computers with next generation voice controls, some of which do not even have keyboards. I will admit, the experience was remarkable. Such computers not only listen to our commands but can interact with them through digital assistants. These assistants use natural language models such as ChatGPT to converse with us while transferring our requests to other AI and non-AI systems for execution.[27]

This transition is occurring at the same time that most consumer electronics are being fitted with IoT capabilities. If AI assistants are given the ability to interact with these devices, we may begin to experience a computer ecosystem more akin to what Weiser imagined, with an AI conductor at the center of our own personal computing symphony. This task is more difficult than one might assume due to privacy and security concerns, as well as the fact that industry standards for connectivity must be established. However, there are numerous groups working on making it a reality.

Theoretically, the more data that is available about the ecosystem user, the more adaptive the system can become, learning our habits and making suggestions, or even automatically completing tasks. Part of what is driving the capability for processing this data is progress in AI computing infrastructure.

Since AlexNet, the industry for producing computer chips that can run AI systems has exploded.[28] Recently, there have been major developments in the production of smaller, cheaper AI-compatible chips,[29] which means that an AI system can be run on a localized device. The data collected by surveilling the user to improve performance can be processed on the user's personal devices rather than in the cloud, which potentially minimizes some privacy concerns.

The ambition to transition to a ubiquitous computing environment, where the computer exists around the user and adapts to them, is not entirely new. Some companies have been very vocal about their ambitions, while others have kept their cards close to their chest while making strategic decisions that suggest they too are moving in that direction.

For a long time, the most vocal company in this space was Google. They have long had a love affair with the idea of ubiquitous computing. Although, it hasn't been as much a Hollywood, candlelight, cheese fondue on a train, hot pink style love affair. It's been more of a two people who met at 2 a.m. in a bar when the lights came on and now do really want to be together but keep breaking up and trying again kind of love affair.

In 2012, Google announced a ubiquitous computing project that is now widely regarded as being ahead of its time. But at the moment, it was just regarded as a disaster.[30] I am of course referring to Google Glass, the first major attempt at smart glasses.[31] Looking back, Glass was a technological marvel. By all accounts, the glasses actually worked quite well and did a good job at mimicking the function of a smartphone. They were a noble attempt at redefining how we interact with computers, making them more reactive to our environments and more intertwined with us as users. The glasses had a series of powerful features, the centerpiece of which was a head-mounted projection display that could display images on the lenses and into the user's field of vision. The glasses also had a front-facing camera, with rudimentary facial and object recognition capabilities, and a voice assistant. I highly recommend watching some recent videos of tech reviewers revisiting Google Glass and its capabilities, as it is shocking just how good the product was. Nonetheless, Glass still failed for a number of reasons.

First, the glasses started at $1,500.[32] By contrast, the iPhone 5, which launched the same year, started at $649.[33] The glasses also raised serious privacy concerns as, at the time, the public had not yet become numb to the idea of everything being potentially recorded all the time.[34] Seeing people walk around with a camera strapped to their face was a major concern. Finally, they might have just been a bit too much, a bit too soon. The glasses had a sci-fi look to them. They just didn't fit in with the aesthetics of the moment. People who wore them were not seen as cool and were often publicly ridiculed[35] – not exactly the best marketing strategy to promote wide-scale adoption. But Google's ambitions did not die with the failure of Google Glass.

In 2014, Google acquired the smart home company Nest for a whopping $3.2 billion.[36] This project has not been a complete failure, as Nest produced a wide array of celebrated smart home products under Google before being completely amalgamated into Google's smart home project/ecosystem.[37] The cornerstone of Google's smart home project right now is their Google Home system, which, admittedly, I do use and love.

In 2017, they revived the Google Glass program, changing its focus to sell to business and industry buyers instead of directly to consumers.[38] But they killed the program again in 2023.[39] The strategy was smart, as it was geared toward addressing problems with consumer adoption. Although consumers may be concerned with how they look when wearing a piece of technology, companies are not. What matters to companies is that the tech works and makes the workers more efficient.

This strategy was mirrored by Microsoft, which released its version of smart glasses in 2016.[40] Although both were glasses, the two products could not have been more different. Instead of a projector that displays 2D images in the user's field of vision, Microsoft pioneered hologram technology that can actually insert an interactive 3D hologram into the user's field of vision.[41] I have used this product myself and can confirm it feels like something out of a science fiction movie. Computer screens and 3D models magically appear all around you in your environment. Part of what makes this possible is an array of sensors in the headset that feed into AI programs, allowing the headset to produce a detailed 3D map of its environment. However, this technology comes at a steep price: almost $3,500.[42]

Although you have probably never seen a HoloLens, they are an extremely popular product. They are very useful in numerous industrial settings such as surgery, manufacturing, engineering, and construction. In fact, the military and Microsoft have been working on a HoloLens contract valued at over $20 billion.[43]

It is likely that Google and Microsoft hoped that the glasses would follow a similar adoption cycle as early computers did, with people first using them at work and realizing their value and potential. After becoming familiar with the products on the company's dollar, a user might decide to buy one for personal use.

I must note that my insistence that Google sees ubiquitous computing as the way forward does not just come from an interpretation of their product launches. They have outright stated their intentions. However, they have attempted to rebrand the philosophy of ubiquitous computing as "ambient computing."[44] In fact, I first learned about ubiquitous computing while investigating Google's ambient computing projects.

Talking at an investors meeting in 2019 about Google's bets on ubiquitous computing, Google CEO Sundar Pichai stated:

> We are very excited by the vision of ambient computing and evolving that. I think it's a continuity in the sense that, over time, computing should be more intuitive to users and computing should adapt to users, not the other way around. And the foundations of all this is all the work we have done with our computing platforms to date and the successful consumer services and the developer platforms we have built. And I think that's the most of the investments there. The phones will continue to be at the center of ambient computing for the future. So that's another important piece where we are already invested in. I think as we expand beyond, and that's what the Made by Google family is focused on, products in your home, with our Nest family of products and wearables, which we do with Wear OS and so on.[45]

Pichai's statement can be put in more context with reference to Rick Osterloh, Google's senior vice president of devices and services. In his keynote speech at the Made by Google conference in 2019, Osterloh said:

> In the mobile era, smartphones changed the world. It's super useful to have a powerful computer everywhere you are. But it's even more useful when computing is anywhere you need it, always available to help … [H]elpful computing can be all around you – ambient computing. Your devices work together with services and AI, so help is anywhere you want it, and it's fluid. The technology just fades into the background when you don't need it. So the devices aren't the center of the system, you are. That's our vision for ambient computing.[46]

It makes logical sense that Google would make such a hard push for ambient computing. They are leaders in AI, and ambient computing is an AI-driven idea. Financially, Google also senses that they need to pivot if they want to survive as a company.

Historically, juggernaut companies like Google struggle to maintain their position at the top of the market. It is a game of adapting to new times or dying. Most major tech companies from the early days of computing, companies that once seemed invincible, have fallen by the wayside. Microsoft managed to avoid such a fate but only because they reinvented themselves as a services company, pivoting away from operating systems and toward their cloud and office suites.

Google and all other surveillance advertising companies, like Meta, are facing massive economic threats right now. Although what they were doing was not illegal, the more the public learned about it, the more pressure mounted on governments to do something. Not only did people not like being watched, but the companies watching us constantly broke our trust. Scandal after scandal emerged, from Cambridge Analytica using Facebook data to attempt to influence elections by directly targeting our unspoken fears and anxieties, to Facebook knowingly ignoring their marketing algorithms' effects on young girls.[47] Around the world, legislation has been passed that fundamentally threatens the business models of these companies. More is probably on the horizon. As such, there is a noticeable scramble, as surveillance companies attempt to find new profit areas. Google is doubling down on AI.

In fact, in 2017, six years prior to the launch of ChatGPT, Google's CEO announced to the company that they would be transitioning to become an AI-first company.[48] This is a sentiment he has repeated numerous times in investor and earnings calls.

Although Google has by far been the loudest player in the ubiquitous computing space, they have not been the most consequential. In fact, the most consequential player is the one that kept things the quietest until they were ready to show all their cards. Although they are not known for their AI development, it is possible that history might once again remember Apple as the company that ushered in the new age of computing.

CHAPTER 18

Apple from the Top Rope

One of my university housemates loved the WWE. Before meeting him, I had never watched it. But when we lived together, it wouldn't be uncommon for it to be playing in the background as I studied. Although I never grew to love it, I did gain a massive appreciation for it. It could be pretty fun. I think the WWE appealed to my love of chaos. When watching normal sports, anything can happen, but at the same time, that's not really true. Sports like hockey or football have a defined set of rules. A basketball team can't just decide one day that they are going to bring pogo sticks on the court with them. Sports have chaos, but it's controlled chaos. The WWE has no such rules. It's entertainment – rules would just hold the wrestlers back. So when you watch, anything can happen.

I especially loved a common bit they would pull where a hated fighter, the villain of the moment, would be wrestling for a championship. Just as it looks like he is about to win, his opponent defeated, the doors to the stadium would swing open. A cloud of smoke fills the venue, and a grizzled WWE veteran emerges – a retired fighter and hero of the sport. Honestly, I had no clue who these people were, but the crowd loved it.

No one knew he would be there. There were rumors, but no one believed them.

Our hero marches to the ring, soaking in his thunderous applause – the trumpets of his miraculous return. He climbs to the top of the ring and raises his arms to hype up the crowd before leaping into the air and landing on his opponent. Even though he missed the first 95 percent of the fight, the audience instantly crowns him the deserving champion.

In our story, it is Apple who emerged from the smoke and leaped from the top rope.

Apple has a long history of being incredibly secretive. Unlike Google, which is often loud and will make bold proclamations, Apple is conservative and reserved. They do not demo technology until they are ready to release it. Their shareholder calls are tightly scripted. And within the company, information is tightly controlled. Project teams are segregated within Apple offices and are kept in the dark about the company's grander strategies. I have even heard reports of employees spending years working on tech with no idea what they are building it for. As a technology researcher, this secrecy can be infuriating. But there is one major window into Apple that can shed some light on what they are working on – their patent filings.

A few years back, I got the tip that Apple was joining the glasses game. And, like clockwork, patents began to appear. But Apple being Apple, there was no certainty to the project. Tim Cook (their CEO) could be ruthless. If the project didn't look like it was going to pay off, I had no doubt he would trash it. Over the last six years, I have heard numerous rumors that the project had been killed or was teetering on the edge. Honestly, I never knew what to believe, and I'm sure the rest of the tech world was in the same boat.

However, Apple slowly started to make moves that signaled something was coming. First, they started laying the foundations for augmented reality applications. They released a toolkit for independent app developers that gave them the ability to create augmented reality apps for the iPhone.[1] Developers making augmented reality apps would be vital for the success of any form of augmented reality glasses. Although it could have been an isolated move, it did seem to hint at something more.

But still, a missing piece for Apple was AI. They had never been a major AI player. Until they were. Suddenly, Apple began releasing fascinating new AI tech. Specifically, they became pioneers in a type of AI training known as federated learning.[2]

Federated learning was first proposed by Google engineers as a more privacy-continuous way to train AI systems on users' personal data.[3] With traditional AI training, the user's data needs to be extracted and brought to a centralized location, where it is mixed with every other user's data for training.[4] This process creates a lot of privacy problems because (1) the company doing the training technically has access to your data; (2) when data is removed from your device, you lose control over it, and it thus becomes vulnerable to data leaks and hacks; (3) there is an undeniable ick factor to a company extracting extremely personal data such as your voice recordings, conversations, biometrics, and so on; and (4) there is a history of these companies being targeted by government agencies who want to spy on foreign and national actors. When data is centralized, it becomes more vulnerable to government interference.[5]

Google's solution was to create a new way of training AI where the user's data never had to leave their device. Instead, an initial AI model is trained on existing non-personal data. That model is then duplicated and sent to devices all over the world, where it trains on the users' personal data. After a period of time, all the individual models that are training on the users' devices return home and are combined to create one supermodel. In this system, the user's data never has to leave their device for the overall model to be trained.

Apple, who have tried to position themselves as a "trustworthy" user of data, took this model and ran with it. In multiple papers I have encountered, Apple researchers have stated their belief that federated learning is the answer for training AI systems at scale, using personal data while respecting users' privacy.[6] Pushing the method even further, Apple has developed a system called federated tuning.[7] It is unclear to me how much AI research Apple was already doing in secret, but around this time, they also began to acquire a lot of AI startups, bolstering their talent pool.

Simultaneously, Apple made massive leaps in the development of AI-compatible computer chips. Specifically, their "M" line of chips shocked the industry when it demonstrated ludicrous computing and GPU capabilities within an ARM architecture.[8] The ARM architecture part is crucial as these are the type of chips typically used in cellphones.[9] They are small and do not produce as much heat, meaning that they do not need a fan. If the chips don't need a fan, and heat isn't an issue, they could be installed in basically any device. The "M" line of chips was a game changer.

Even as all this was happening, it was unclear what exactly Apple's plan was. Apparently the same was true within Apple.[10] Tim Cook had founded the glasses program and reportedly saw it as his iPhone moment. But the actual glasses technology was moving slower than anticipated. However, there was another product, within the same family, that was showing promise. It was a device similar to a virtual reality headset but that acted like the desired glasses. This is the device we now know as the Apple Vision Pro.[11] The decision to release the Apple Vision Pro was, according to a report by the *Financial Times*, a contentious one within the company.[12] Designers wanted to wait until they had perfected the technology and could release a model that better resembled a traditional pair of glasses. But in the end, Tim Cook decided it was time to announce Apple's entry into ubiquitous computing to the world.[13]

A ski-goggle–like headset, the Vision Pro has an impressive number of high-tech cameras and sensors attached to its sleek design. The sensors allow it to construct a detailed mapping of both the environment around it and the person wearing it. The information captured by the sensors is then relayed back in real time to a virtual reality video screen, recreating the outside world for the wearer. This technology is called passthrough, because, even though it is technically virtual reality, it is practically doing something quite different. It is trying to mimic the effect of clear glasses. Apple Vision Pro is not the first product to try and do this. But it is the first to do it at this level.

I have personally used an Apple Vision Pro, and the virtual recreation of reality is damn good. At times, you forget you are in a simulated environment. This effect is heightened as the array of sensors removes the need for a remote. Instead, your hands are tracked and displayed in the virtual reconstruction.

Within the virtual reconstruction of reality, the impossible becomes possible. 3D models can be dropped into your surroundings. Computer screens can appear anywhere you want them. The world really does become a computer. And more, the headset is embedded with impressive AI features that allow it to predict your actions and expectations, and adapt your experience to them. This is made possible by its unthinkable collection of biometric sensors, which are so advanced that they can collect information about your eye movement, electrical activity in the brain, perspiration, muscle contractions, and heartbeats. One Apple patent referencing the device states:

> Some implementations disclosed herein present a computer-generated reality (CGR) environment in which a user participates in an activity, identify a cognitive state of the user (e.g., working, learning, resting, etc.) based on data regarding the user's body

(e.g., facial expressions, hand movements, physiological data, etc.),
and update the environment with a surrounding environment that
promotes a cognitive state of the user for the activity.[14]

Like Google, Apple has attempted to rebrand the concept of
ubiquitous computing so that they may be the ones associated with
its development. Specifically, they have renamed it "spatial computing." When launching the Vision Pro, Tim Cook proclaimed
to the audience, "Today marks the beginning of a new era for
computing." The Apple press release noted, "Just as the Mac introduced us to personal computing, and iPhone introduced us to
mobile computing, Apple Vision Pro introduces us to spatial computing": a pretty clear statement of Apple's objectives.[15]

Apple Vision Pro was not the first product of its kind to be launched.
But there was something unique about its entrance into the space.
Before the Vision Pro, no one seemed to care about my obsession
with studying wearable technology and the philosophy behind it.
Then suddenly, people started asking me about it. Everyone had
heard about the Vision Pro, and as it was Apple, everyone was inter-
ested. In my opinion, Vision Pro did for ubiquitous computing what
ChatGPT did for AI. It made it more visible to the general public.

Since its release, the Apple Vision Pro has been met with glow-
ing reviews. Reviewer after reviewer has raved about the technology
and made statements about how it's like "looking into the future
of computing," that spatial computing is the future of human–
computer interaction. The problem is, Apple Vision Pro is not
very accessible. It costs $3,500, and Apple only expected to sell
about 1 million in the first year.[*16] There also don't seem to be
many practical uses for it yet, making it a very expensive toy.

---

* Due to manufacturing and supply chain issues, Apple eventually scaled back their sales
forecast to 400,000 units.

But I think what Apple recognizes is that they are at the beginning of something new and exciting. And they are trying to mark their territory. The original iPhone only sold about 1 million units in its first year, far fewer than you may have thought given its reputation now. It's about introducing their idea of spatial computing to the masses, getting the public excited, and getting developers on board with making new programs for the technology. Apple's next task is to perfect the technology and make it more consumer friendly and cheaper, ideally leading to mass user adoption.[17]

If they do accomplish the goal of creating a traditional glasses-like product, it is possible that we may remember the Apple Vision Pro in the same way we do the Apple Lisa: an expensive stepping pad to the future of computing.

***

Although the Apple Vision Pro made a big splash, it is not the only player in the game. I need to give credit where credit is due. Meta have similar, and much cheaper, passthrough products and have invested heavily in the virtual reality industry. However, I think the most interesting product Meta has in their portfolio is their Ray-Ban Smart Glasses.[18]

Meta's smart glasses are quite humble. They are very stylish and have only very basic features when compared to Apple's. They have a front-facing camera and an impressive audio function that replaces the need for headphones. Really, the only thing that is actually "smart" about the glasses is an AI feature that allows you to converse with an AI assistant, which can observe your environment through the front-facing camera. So, if a car drove by, I could ask, "Hey Meta, what type of car was that?" And the glasses would answer my question.

As of now, there are no augmented reality features. However, in a company-wide meeting held in early 2023, Mark Zuckerberg reportedly presented a plan to release augmented reality features in 2025 and fully capable augmented reality glasses by 2027.[19] Zuckerberg has also reportedly stated that he strongly believes that this technology will replace the mobile phone within his lifetime.

On September 25, 2024, Zuckerberg publicly demoed Meta's prototype augmented reality glasses for the first time.[20] Although the glasses are not yet ready to be sold on shelves, they were further along than many had assumed, and Meta presented a clear path to making the technology consumer ready.

This movement is not constrained to big tech or even just to the idea of glasses. Smaller companies such as Brilliant Labs have been working on cheaper, open source alternatives to big tech smart glasses.[21] Companies like Humane have pitched ideas such as replacing the phone with a chest-mounted, AI-powered laser projector.[22] And Naqi Logix has designed earbuds that can allow humans to control computers with microgestures.[23]

The point of this section has not been to speculate on what the future of computing will look like or to promote specific visions of ubiquitous computing. Instead, what I hoped to show you is that there is a powerful current in the tech industry right now, driven by the philosophy and belief that ubiquitous computing is the future. Whether the project as a whole will succeed is unknown. However, it is undeniable that there is a wave coming. And this wave is being pushed by the rise of artificial intelligence.

AI has allowed computing devices to have eyes, to view the environment outside of their circuits and try to make sense of it. It also allows the machines to judge us, to take the information from biometric sensors and try to come to conclusions about our unspoken thoughts and feelings. Moreover, it has gained the ability to relate our bodily data to the data it is collecting about the world around us.

Ubiquitous computing is a complete melding of man and machine. Through sensors, the computer extends through our body and into the world. It becomes part of us, reacting to our body and editing the world around us, as if it were an organ in our body. We are becoming cyborgs of sorts. Moreover, it is hard to actually pinpoint what exactly the machine is. Increasingly, the machine is ceasing to be a single object in our world as smart computing technology becomes more prevalent.

Interestingly, although the IoT was originally imagined around a wheel and spoke model, where there would be a central conductor, such as a single AI system, used to process the information of each device, the grand vision has changed. Now what is imagined is an IoT mesh, where each device can have its own embedded computer, and each device in the system can communicate with every other device: a web of computers, a completely interconnected system.

The world is becoming more and more digital. Computers are becoming more deeply embedded into every part of our lives. Even objects that don't have computer chips in them are still being integrated into the digital through the use of smart cameras that can watch and analyze them.

The more that is digitized, the more data there is.

The project of ubiquitous computing has become a project of making the world more legible to computers, so that computers can change the world around us – potentially without us even asking. This may come with some serious unintended side effects, some of which are obvious and some less so.

CHAPTER 19

The Surveillance State

The East German Stasi are one of the most infamous, and nefarious, spy organizations ever to exist.[1] Their mission was to know everything that went on in East Germany during Soviet occupation. If two people slept together, they wanted to know. If you had a dissenting thought while taking a bath, they wanted to know. They were ubiquitous. When the Soviet Union fell, the world got a chance to see just how intense the surveillance had been.[2] An audit of the Stasi's files showed that, at their peak, they had an open file on one in every three citizens of the nation.[3] Collecting this information took unthinkable manpower and dedication. They are remembered as monsters.

Contemporary mass surveillance is much more benign. As surveillance scholar van Otterlo notes, most of society now happily does the job of a Stasi agent without hesitation.[4] Through social media, we neatly assemble a file of our age, address, and contact information, paired with a list of our contacts, relationships, and personal interests. In the age of social media, surveillance has never been easier.

The two are not direct comparisons. As sinister as companies like Facebook may seem at times, they are not the Stasi. But the undeniable fact remains that we live in the golden age of surveillance.

It's an uncomfortable and sometimes inconvenient truth that not all surveillance is bad. Surveillance can keep us safe. It can help us better understand and organize ourselves, and it can be used to keep the powerful in check.[5] The media is supposed to be a surveillance tool that keeps governments in check. But it can also be used for evil: to control populations, to manipulate us, to chill free speech, and to squash dissent.

It is a fine line, and it is easy to accidentally cross over from well-meaning surveillance to oppressive surveillance. There is an entire academic discipline (to which I contribute) called surveillance studies dedicated to researching the effects and practices of surveillance. It is our job to try to understand when surveillance is being abused and when it is harming society more than it is helping. And let me tell you, things have become seriously complicated in the past twenty years.

The digitization of society has greatly increased the amount of surveillance we are under and altered who is conducting mass surveillance. In the past, mass surveillance was mainly conducted by governments who were interested in controlling the population. Again, not all control is bad. A non-authoritarian government is still likely to surveil the population to protect civilians and to control threats such as plagues, famines, or terrorism. Surveillance is also vitally important for everyone's least favorite government activity: taxes.[6]

We are often blissfully unaware of how integrated tools of surveillance are in our everyday lives – specifically, tools of legibility. Legibility tools are not the technologies that watch us but the technologies that make it easier for the watcher to understand, record, and manipulate us. Legibility tools make the messy world that one may want to surveil artificially more ordered.[7] Unknown to you,

you have almost certainly carried one of these tools with you since your day of birth: your last name.

Humans have long used names to identify one another. They are a technology so old that we have no clue when they were invented. Last names, however, have a much more checkered history.

Last names emerged in different civilizations at different times but always for similar reasons.[8] In small social groups, there is no need for last names. Everyone can know everyone. If there were two "Peters," one might be referred to in reference to their father: Peter John's son. But such a practice was quite crude and not official. Last names were often mandated as a way to identify and differentiate between individuals for social or governmental reasons.[9] Sometimes last names were assigned in order to differentiate the powerful from the powerless.[10] Royals and their aristocracy were given special last names. This practice made the members of these families more legible, as all you needed was their last name and you knew a lot about the person. Information was tagged to the name that was passed between family members. But by and large, the most common reason for adopting last names was taxes.[11]

Taxes are a practice that require a massive level of surveillance. If you are charged with collecting taxes for a given area, you need to know a lot of information. Who are you in charge of? What do they do, and where do they live? And most importantly, have they paid you? It would be impossible to keep all this information sorted in your head, so you need to record it in a table. Already a difficult task, this job gets even harder when there are 100 people called John in your community. But there is a simple solution. You make everyone adopt a second name, a last name, which makes them more identifiable. In fact, why don't we add a middle name to make each person's name more unique?

This was the exact problem encountered by Narciso Clavería y Zaldúa, a former Spanish army officer who served as governor of the occupied Philippines from 1844 to 1849. One of his main tasks was collecting taxes. But he found this job difficult because the people of the Philippines used a complicated and non-standard naming system, making them difficult to keep track of. So he compiled a list of every Spanish last name he could think of and then had his subordinates take the list to each village and assign last names to each inhabitant. In order to make the population even more legible, names were given alphabetically, based on where the person lived. Everyone in the town of Oas had a name that started with the letter "R." The same went for "G" in Guimbal, and so on. If you meet someone from the Philippines, it is possible to pinpoint exactly where their ancestors lived at that moment in history, based on the first letter of their last name.[12]

Technologies of legibility are important because surveillance is always conducted with a goal. In this case, it was to collect taxes. You can theoretically collect taxes without last names, but it is more difficult. Making a population more legible simplifies the world and makes it easier to process.

Increasingly, over time, more and more mass surveillance has been conducted by private companies. One of the first major private industries to conduct mass surveillance was the insurance industry.[13] Companies used the information they gained, processed through mathematical algorithms, to decide how much to charge for coverage.[14] Private mass surveillance began expanding even more as finance became more privatized.[15] The rise of the consumer credit industry in the mid-1900s created the need for private companies to know an increasing amount about individuals in order to determine their trustworthiness with regard to paying their debts.[16] This need expanded when credit card companies realized that they could make more money from people who wouldn't default, but couldn't, or didn't, fully pay off their debts, resulting in higher interest payments.[17] If a company can determine that you are in this group, then they can more aggressively advertise credit cards to you.[18]

Advertising is also an industry heavily associated with surveillance. Advertisers want to know who they are marketing to so they can be more efficient.[19] As such, smaller surveillance industries developed with the goal of acquiring consumer information that could be sold to interested advertisers. Prior to the internet, it wouldn't be uncommon for marketers to buy lists of potential buyers' addresses and phone numbers to directly market to them.[20]

This form of surveillance accelerated with the computer and internet revolution and the rise of targeted digital advertising.[21] As we discussed in chapter 10, surveillance advertising created an economic justification for collecting and processing the information created by our digital presence.

Surveillance also increased as we developed more technologies to process digital information. Contrary to popular belief, surveillance doesn't just affect humans. It can be directed at all sorts of different actors. We surveil the atmosphere, collecting data on all sorts of variables such as temperature and precipitation in order to try to predict the weather. Financial analysts surveil financial performance, and other company indicators, in order to try to predict the stock market. Through the act of surveillance, we create data that we can process to attempt to achieve a goal.

In the age of AI, there is a new reason to conduct surveillance. As we have explored in depth, deep learning technology is built on the foundation of data.[22] Data is the fuel that powers the engines of the AI revolution in the same way that fossil fuels like coal and oil fed the engines of the Industrial Revolution. And data is the product of surveillance.

Data, at its core, is a collection of perceived facts. To create data, you need to surveil a subject and record its actions in a legible format. In the case of AI, a legible format is one that can be processed by a machine to help it achieve its goal.

Photos taken on a Polaroid camera are data; however, in the form of physical film, they are not good data for machine learning. A machine can't process film. However, if we scan the photo into the computer, we can transfer the image into a format the machine can understand, a collection of 0s and 1s that come together to create a representation of the image. This is data that the machine can process.

AI has also helped us make data more easily legible to machines. When AlexNet was first programmed, the images were digital, but the machine needed data regarding what was in the image in order to train.[23] Humans had to go through each photo in the dataset and individually label them in order to make the images more legible to the machine.[24] Now that we have AI systems that can recognize objects in images, we have automated the legibility process. A camera recording a subway station can now recognize and record approximately how many people walked through the station that day. This data can then be used to train more AI systems.

The development of AI systems that can understand and translate video and audio has made the everyday world far more legible to machines. On top of that, we are installing ever more sensors all over the place: sensors that detect your heart rate, sensors that track eye movements. Data-producing surveillance is everywhere: in your pocket through your cell phone, on the street through sensors and cameras, and on the internet through a massive array of web tracking technologies. OpenAI uses a surveillance tool known as a web scraper to record the text written online, producing training data for their AI.[25]

The generation of data is a massive industry, but to do it, companies need you to enter environments that are built for surveillance. The more digital the space, the easier it is to produce data.

As such, companies create what my favorite surveillance scholar, Mark Andrejevic, calls digital enclosures.[26] These are spaces where they can monopolize the act of watching you and transform your actions into data.[27]

Google Search is a digital enclosure.

The Apple ecosystem of devices is a digital enclosure.

Facebook is a digital enclosure. The concept of the Metaverse is a space where every action taken becomes data – the ultimate digital enclosure.[28]

When we examine the trend of ubiquitous computing, we can observe a massive expansion of the surveillance apparatus. When the computer exists all around us, more and more of our physical and social world become translated into data to fuel AI systems.

Discussions of data privacy usually revolve around the effects of targeted advertising. Most data privacy bills focus on what is called personal data. It is data that is directly linked to the person it is collected from. Control over data given to citizens by legislation such as the European Union's General Data Protection Regulation is limited to personal data.[29] This makes sense as such legislation was made in response to the harms of digital advertising. But AI is a different beast.

AI companies are becoming less interested in data they can link to an individual person because they are not interested in marketing to the individual. Instead, what they need is massive amounts of general data, which can be used to train AI systems.

One of the major reasons I decided I needed to write this book was because I know that the general population is unaware of just how big the surveillance apparatus, which is built to feed AI, is growing. It is a unique type of surveillance system compared to any we have seen before. These companies are not the Stasi. They are not inherently collecting our data to oppress us. But that doesn't mean it's okay. It is a massive societal shift, and we are writing the rules for how to deal with it as we speak. I think you deserve to know how we are building the surveillance infrastructure that makes AI possible.

When the Industrial Revolution occurred, the changes to society did not just occur in the factory. The changes were not just how we worked or the types of products we could make. The invention of the steam engine had far-reaching social effects that reshaped the make-up of how we lived. It changed the way society as a whole functioned based on the activities that were necessary to make the engine turn and the opportunities that fueling the engine created.

As a scholar of AI, it is not my job just to understand how AI works – it is my job to try to understand the deeper implications and to share them with you. So, buckle up, because things are about to get weird.

CHAPTER 20

Data and Power

It has become a common saying that data is the new oil. I hear it everywhere I go.[1] Even I said it in the last chapter. The thing is, it's not really true. Data and oil are nothing alike.

Okay, they are similar in one major way. Where oil was used to power engines, data is used to power AI. But that's really where the resemblance stops. And, as a lot of people far smarter than me have pointed out, this comparison conjures up a disingenuous image of where data comes from. But first, let's talk about oil.

Oil is found in the earth. It existed before our civilizations, and we found it. In order to harness its power, we first need to dig it up. Then we need to refine it. And then it can be shipped around the world to be burned or turned into plastic or gasoline.

We have fought a lot of wars over oil. A lot of blood has been shed to get it and to defend it. That is because oil is what economists call "a rival good."[2] That's a fancy way of saying that only one person can use it. If I have a barrel of oil, only I can burn it. And when I do, it's gone. Oil is also what we call a "fungible commodity," meaning that every barrel of oil is more or less the same.[3]

Practically, it doesn't matter where your barrel of oil came from since all oil burns the same. But none of that is true for data.

Data is not a fungible commodity as each piece of data is different. In fact, data being varied is what makes it valuable. You can't train an AI system to recognize cats with a dataset of images of a single cat.

Data is also non-rival, meaning that more than one person can use the exact same data.[4] If, instead of a barrel of oil, I have a hard drive full of data, when I use that data to train an AI system, the data doesn't vanish. It doesn't even change. I can then take that hard drive and give it to a friend to make their own AI system. I could also just take that data and email it as a packet all over the world, allowing an infinite number of people to use it to train AIs. That is essentially what Fei-Fei Li did with ImageNet when she made it open source.[5]

Finally, data isn't just found in the ground. It is produced.[6] It is made through the act of surveillance. Oil has no attachments. Once a government awards drilling rights, the oil company technically owns that oil. But that's not true with data, especially data produced by surveilling humans.

The question of data ownership is extremely complex and is the subject of fierce debate on the pages of law journals around the world. And I will be honest, this conversation is a very complex one at the cutting edge of research right now. Exploring this topic properly, and with full nuance, would be an entirely different book.

However, an important thing to note is that, as of right now, it is generally accepted that one cannot own data.[7] It cannot be property. Yet there is still a massive market in it because, as my colleague Teresa Scassa elegantly points out, a massive market has emerged that buys and sells data based on restrictive contracts.[8] That sentence probably made little to no sense, but it is extremely important, so let's dig into it.

When data is created through surveillance, the relationship between the watcher and the watched is not equal. The watcher has the technology that is recording the watched, whether it is with a pen and paper or a complex data-scraping algorithm. Because they have the technology, they possess the data itself at the end of the interaction. Even if they do not legally own the data, they are the ones in possession of it. They can then choose to provide other businesses with access to this data. What they are selling is not the data itself but access to the data.[9] When negotiating this deal, the contract will stipulate that the person buying access to the data cannot share it with anyone else.

In this situation, the person who conducted the surveillance does not own the data but is able to act as if they do because they are the only ones with unimpeded access to the data. Even in systems such as the European Union, where the surveilled person is legally entitled to request their data from a surveilling actor such as Microsoft or Amazon, the data is basically useless on its own for the purpose of AI training.[10] What is valuable is the collection of thousands of people's unique data to which the surveiller retains exclusive access.

This system awards a massive amount of power to companies that already have access to a lot of data and have, or are developing, massive surveillance infrastructures.[11] Those companies thus gain the ability to control the flows of data throughout society.[12] They get to choose who gets access to it to innovate and who does not. Although data protection laws may restrict some of their ambitions, such laws do not apply to the vast majority of data. Data that is not considered personal is exempt from this legislation. Interestingly, data you may believe is covered by legislation probably is not. Although you may consider your heart rate personal information, under many jurisdictions, if the company operating your smartwatch sells it in an anonymized form – simply meaning they have removed identifiable markers such as the name attached to the file – they can freely sell access to this data or use it however they would like.

This system is very different from its oil-selling predecessors, partly because data as a good can be sold an infinite number of times. There is no limit to how many people can use the data once the surveillance company produces it.

The topic of data economics is incredibly complex and highly popular right now. This interest is because data as an economic good works so differently from previous primary goods such as oil. Because of that, governments need to rethink their economic rules to govern how data flows throughout a given economy. These systems are rapidly developing and will influence how AI is developed and who controls it within different countries and regions.

For example, as reported by one of my closest colleagues, Brett Aho, China has taken an approach that he calls "data communism."[13] The Chinese government has declared data to be a "factor of production," meaning that it belongs to the government and not to the entity that produces it through surveillance.[14] This approach means that the government can provide access to the data to any interested party for the purpose of training AI systems. This approach is likely to give Chinese companies a massive advantage when it comes to making AI systems. However, it comes at the cost of the Chinese people's right to privacy and security.[15]

On the other hand, the United States currently has an extremely anarchist capitalist system. In this system, companies control the data they produce through surveillance and get to decide how it is distributed. Some groups decide to make non-personal data open source. These groups believe that the more data that is available, the better systems we will make, and thus all of society will advance. However, in the capitalist system, most data is kept under lock and key, and the companies that control it are currently the most valuable companies in the world.

Notably this system also has its privacy problems. Innocent people's data is produced and sold without their knowledge or consent. Furthermore, there are major questions that need to be asked regarding how we feel about these companies gaining so much power through the surveillance of society. Although their surveillance technology is vital for the production of the data, it is only one-half of the equation. They are benefiting from recording us. A self-driving car doesn't just produce data about its own driving but also records all the other drivers on the road, as well as the behavior of every person walking through the public square it drives through. In attempts to remedy this situation, new proposals for how data economies can function have emerged, such as digital trusts and data commons.[16] There are also proposals for systems where individuals own their data and can then sell it back to the surveillance companies.[17]

Each of these systems has its own problems. There is no perfect solution. Whatever we decide to do, there will be trade-offs. Some systems value privacy above all else, but that comes at the cost of innovation. Others value a more equitable sharing of data resources but recognize that doing so may reduce investment in data production operations. It is a tricky balance. I spend a lot of my time working on this problem, trying to develop tools that can help governments make more informed decisions about how the choices they make when constructing the rules for data economies will affect the types of AI developed and the way their societies function.

However, the big thing that I want to highlight here is power. Those who control data are granted immense economic and social power. They get to decide what types of AI are made, and for whom. They decide what types of problems we try to solve, and how. My greatest fear is that, right now, many countries, including my own, are ceding too much power over how our futures will be shaped to companies whose motives are not to make a better society for all – but instead to accumulate more money and power.

CHAPTER 21

The Ubiquitous Machine

Ubiquitous computing has the potential to be a society-altering technology. But I must again enforce the idea that no technology is apolitical.[1] We do not simply build a technology as a natural product with a predefined destiny for how it will affect society. When we build tools and systems, we make choices. Those choices may be subconscious or unspoken, but they are choices. And they reflect a specific ideology of the world we wish to build.

What I am trying to say is, not all visions of ubiquitous computing are the same.

Systems such as ubiquitous computing give the illusion of personalization, but it is just an illusion. The act of personalization happens in a defined box. The Apple Vision Pro might change the world around you to fit what it thinks you want, but the changes it makes are still constrained by the preprogrammed nature of the device. There is an operating system that it must work within and potentially instructions regarding what it can and cannot do. In short, those who design the ubiquitous systems we interact with subtly, and not so subtly, control the way we understand the world we live in.

Take Google Search for example. When I Google "pizza places near me," I may feel empowered by the countless options Google gives me. But am I? To be honest, I usually only look at the top one or two options. Although I may have the illusion of choice, I am far more likely to pick one of the first results. Which results come up first affects the way I live my life because it affects what pizza I am going to eat. And Google knows this; that's why, when I Google "pizza," the first result is a paid advertisement from Pizza Hut made to look like a search result.[2] Although Google's mission is to provide me with information, they are still a business that wants to make money. And even if they didn't put that Pizza Hut advertisement at the top of the page, they still designed the algorithm that decides which pizza places appear at the top. Their decisions on which variables decide what pizza place should appear at the top will affect my pizza eating habits.

On YouTube, the algorithm is god. YouTube superstars are made and broken by the algorithm.[3] Although I may feel as if the algorithm is tailored to me, it is also tailored to favor specific types of content. The algorithm is designed to favor specific genres and lengths of content depending on YouTube's growth strategy.[4] YouTubers (who on paper have complete control over what they post on the site) rely on securing a spot in our feeds or on our home screens.[5] To do this, they often need to change their content to match what YouTube is favoring at any given moment. The decisions of what type of content to favor are decisions made by YouTube.[6] They are probably made for business reasons in an attempt to grow the website's reach, compete in new markets, or to appeal to advertisers.

Technology does not just exist. It is designed. And design choices are influenced by the humans behind them. It could be one human, as is the case with Elon Musk at the website formally known as Twitter. Or it could be a corporate ideology mixed with executive decisions, like at Apple.

This topic is a crucial one to think about if ubiquitous computing is to be the next wave of human–computer interaction. Although computing companies have always to some extent controlled the way we interact with the world, this technology brings a whole new level. With ubiquitous computing, the line between the computer and our realities is blurred to a whole new level. If I am walking through the grocery store wearing my smart glasses and ask my AI assistant, "Is this healthy?," it might give me a simple yes or no answer. Because of its answer, my experience of the world is influenced by my AI's idea of what is healthy.[7]

These effects can vary greatly in size. They may be small, like the example of the grocery store, or have major effects, like influencing the way we design our cities.[8]

The example of how we design our cities is not hypothetical but is instead something that we are already experiencing.[9] As I mentioned before, smart cities are an example of ubiquitous computing in practice; by examining them, we can already see how the ideology of different controllers affects how people experience the ubiquitous computing environment.[10]

In 2020, a research group composed of Monique Mann, Peta Mitchell, Marcus Foth, and Irina Anastasiu published a brilliant case study exploring the difference between Barcelona's smart city, which is owned and operated by the city government, and the failed Google Smart City project in Toronto.[11] They explore what is seen as the fatal flaw in Google's proposal that eventually led to a massive public backlash and the project's eventual cancellation. Smart cities are supposed to improve the city for its inhabitants. It is the job of a local government, whether conservative or liberal, to try to improve the city for those who live there. That is the mandate of the government. But as a private company, that is not your end goal. Google's values may have aligned with the needs of Toronto's citizens at times, but not always. Google needs to serve their own self-interest, surviving as a company and making a profit.[12] Because of this need, the design for the city takes on a different form. It is made to look futuristic and flashy. The way the city was to be built and the smart tech to be used would be in line with Google's idea of what was best for the city. But that was not necessarily the same as the people's, even if it might provide some benefits.

By contrast, Barcelona's smart city is a public project, and the citizens of the city have a direct say in proposing how it should be used to improve the city.[13] The data produced by the city is, when appropriate, released as free datasets that its citizens can use for their own projects or to propose new ways to improve the city.[14]

What I find most fascinating about the Barcelona project is that, when you are in the city, you would never guess that you are in a smart city, let alone one of the most advanced and successful smart cities in the world. The organic development of the smart city in Barcelona is actually much more in line with Weiser's original vision of ubiquitous computing.[15] The computers are everywhere, but you don't really notice them. They change the environment but in ways that are subtle. Instead of dominating the world, they enhance it. Barcelona's smart city infrastructure has been used to determine the best location in which to plant new trees, improving the city's climate control, bringing down the temperature in the summer, and helping to reduce noise pollution.[16] It has also been used to redesign the city's traffic infrastructure, creating what is now known as the superblock system.[17] This system is a city layout that makes the city far more walkable, increasing citizen happiness and local commerce while still allowing convenient car access. The city also has a smart power grid, traffic system, and public transit scheduling, plus a tapestry of other smart systems running at any given moment.[18]

However, I don't mean to say that we should just give the surveillance apparatus over to the government. In fact, even while writing this chapter, I could hear some of my readers moan when I stated that it is the mandate of governments to make life better for their citizens. It is true that this is technically their job, but in practice, things can be quite a bit more complicated.

Companies are not the only ones with ideological beliefs. AI surveillance in the hands of governments, even the "good guys," can have deadly repercussions. Just ask Alan Turing. He fought for the good guys in the Second World War – and then the "good guys" found out he was gay.

CHAPTER 22

The Apple

In 1952, the Manchester police discovered that the famous computer science professor Alan Turing was in a sexual relationship with another man.[1] His history as a war hero was still classified at the time, but it is unlikely that even that would have saved him.[2] Homosexuality was still illegal, and the British public was still decades away from accepting it.[3] He was publicly ousted, criminally charged with indecency, unceremoniously stripped of his security clearance, and banned from his research.[4] Potentially worst of all, the government injected him with chemicals to chemically castrate him and "de-feminize" his body.[5]

Two years later, the burden proved too much for the young Dr. Turing. As the story goes, on June 7, 1954, he injected an apple with cyanide.[6] With one bite, Turing's spirit parted ways with his body. He was only forty-one years old.

Humans can be a cruel, terrible species. We have so much potential to build together and to thrive. But our potential is poignantly mixed with our tendencies toward cruelty and destruction. At the end of the day, we are just a bunch of territorial, monkey-brained primates who learned to use tools and make fire, and now have to find a way to live on a planet with 8 billion of us – and the internet.

We are also storytellers. We love to tell stories that make us feel better about ourselves. Sometimes, we even give these fictions the title of "history." It can be hard to admit that some of those stories aren't as true as we'd like. And that sometimes, even if we truly are the good guys in some stories, that doesn't mean we can't be the bad guys in others. The world is complicated, and two things can be true at once.

A common example of how computing and data can be used for evil is Nazi Germany's partnership with IBM. Edwin Black wrote the definitive book on the subject, *IBM and the Holocaust: The Strategic Alliance between Nazi Germany and America's Most Powerful Corporation.*[7] It's really not a great look for the legendary American company, but it also shows the devastating results of combining computing technology, surveillance, and a genocidal totalitarian government.

Before I switched my research focus to AI, my area of expertise was social movements. I was particularly interested in their dark side. Like many other researchers, I was trying to figure out why people would follow leaders who were clearly evil. I wondered how it was possible for entire communities to become so filled with hate that people would commit unthinkable acts against their neighbors, acts like genocide. A lot of my time was spent learning about Germany during the lead-up to the Second World War. However, in 2019, my work went beyond just reading.

In the spring of that year, I traveled to Germany and Poland for eleven days with a group of young scholars to study the sites of the Holocaust. We met with academics and survivors, eye-witnesses to the terror. I stood in the room where the genocide was planned and walked the camps where it was executed. That trip changed me as a person. Among the many powerful statements shared during this trip, one in particular has resonated with me over the years. I can't recall who said it or exactly where we were. Even so, it left an indelible mark.

It went something like this:

People come here and vow "never again." Yet history repeats itself. It keeps happening again. There have been over twenty genocides since the Holocaust, and there are bound to be more. It seems it's in our nature. I hope we can stop them from happening. But at the very least, we need to make sure one like this never happens again. And we have been decent at that so far.

The Holocaust was uniquely appalling because it represents the first time an industrialized nation committed genocide on an industrial scale. They had a level of planning and execution that was horrifyingly precise.[8] Yet even the Nazis were limited by the technology of their time. They couldn't find everyone.[9] And they had to slow the killing due to limitations on the number of bodies they could burn per day. The industrial nations of today probably wouldn't have those problems.

All genocides are an affront to humanity – a stark violation of human rights. But there is something uniquely horrifying about the concept of an industrial genocide.

To make their crimes more efficient, the Nazis solicited the aid of one of the leading technology companies of the time, IBM.[10] Before the age of general computers, IBM produced tabulating machines.[11] These were specially designed devices for the expedient processing of information. The tabulating machines were inspired by an old friend, the Jacquard Loom, and ran on punch cards. But there was a significant difference; instead of the punch cards being used to program the device, they were used to store and sort data.

The primary use of this technology was to conduct censuses.[12] A census worker could travel door to door and question each resident. Each resident would be assigned their own punch card, and their answers would then be recorded by punching little holes in the card. You could record their location as well as any features of the respondent: their age, gender, height, religion. All the cards would then be fed into a tabulating machine that, when programmed, could answer questions about the population. It could tell you exactly how many males of fighting age lived in a given area. Or the exact location of every Jewish resident.

When the Nazis first bought IBM's technology, the war had not yet started, but their desire to purge the nation of those they deemed undesirable was clear. Yet IBM still sold them the technology and provided training and aid.[13]

After the war started, IBM did not end their relationship with the Nazis, but acted as willing collaborators.[14] The first thing the Nazis would do after conquering any new territory was conduct a detailed census using IBM's technology. Sadly, the company's involvement went far beyond just helping the Nazis locate their victims. IBM, through their German subsidiary, sold and installed data processing machines in the Nazi death camps.[15] They worked directly with the regime to create custom punch cards that aided the regime in operating the camps more efficiently.[16] In fact, the president of IBM even personally paid to have a bunker built at one of the camps to protect the machines from Allied bombs.[17]

The story of IBM's collaboration with the Nazis is a powerful story of the horrors that totalitarian governments can commit when given access to advanced data processing technology. But it is not the full story.

IBM didn't just work with the Nazis. They also worked with the Americans.

I want to be clear. The Nazis were one of the most evil regimes in human history. There is no crime that can be compared to the Holocaust. In the grand narrative of the war, the Allies were the good guys. But that doesn't mean we didn't do terrible things as well.

After the bombing of Pearl Harbor, the government of the United States rounded up and imprisoned residents with Japanese heritage.[18] Over 100,000 people were extra-judicially arrested and held in camps.[19] Both historians and legal experts have since referred to the camps as "concentration camps."[20] The US president at the time was known to use the term as well.[21] The majority of the prisoners were American citizens.[22] In California, anyone with any more than one-sixteenth Japanese blood was considered a threat and sent to a camp.[23]

Detainees were allowed to bring with them only what they could carry. The lucky ones were able to sell most of their belongings before they were interned. But the majority lost everything: their property, their businesses, and their family heirlooms.[24] All taken away because of the crime of having Japanese blood in America.

These were no day camps. They were located in harsh and unforgiving environments.[25] Inside the camps, detainees were put to work. They engaged in agriculture, manufacturing, and medicine. And some detainees were even allowed to leave the camp to attend college or serve in the armed forces.

Although you may have already guessed that the IBM machines were vital for identifying and rounding up the Japanese, that was not the only way they were used. Just like in the German death camps, the punch card system was used in everyday operations.[26] Most notoriously, they were used to try to classify and segregate prisoners.[*27]

---

* At the 2023 4S conference in Honolulu, I attended a brilliant presentation by Dr. Clare Kim on the "Yellow Peril" during the Second World War and the use of IBM tabulators in US-operated internment camps. Although Dr. Kim's research has yet to be published, her presentation and research motivated and informed my exploration of the topic.

Detainees were given a questionnaire designed to test their loyalty to the United States.[28] Using the tabulators, risk assessments
were assigned to each camp member, determining what type of
stay they would experience. Those whom the tabulator deemed
worthy were allowed to travel outside the camp to go to college or
to work. Thousands were deemed to be "unloyal" and were classified as "No-Nos." The No-Nos were sent to a special camp named
Tule Lake, an especially harsh place.[29] After the war, the tag of No-
Nos did not fade, and the inhabitants of Tule Lake faced extreme
discrimination due to their perceived disloyalty.

Notably, not everyone in Tule Lake was a No-No. The No-Nos
were a specific group that had answered "no" to two specific questions on the questionnaire:

– Question 27: *Are you willing to serve in the armed forces of the
  United States on combat duty, wherever ordered?*
– Question 28: *Will you swear unqualified allegiances to the United
  States of America and faithfully defend the United States from any
  and all attack by foreign or domestic forces, and forswear any form
  of allegiance or obedience to the Japanese emperor, or other foreign
  government, power or organization?*[30]

In these questions, we can observe a major problem with classify-
ing people based on data. Data is seen as being objective. But it is
not. Like technology, there is a human choice about which data
is collected and how it is interpreted. The No-Nos answered "no"
to both of these questions. Yes, that is a fact. But the data smooths
over the complex realities. For example, many No-Nos feared that
answering "yes" to these questions would be considered akin to
volunteering for military service.

It also can't be ignored that they were being asked to pledge
their allegiance to a government that had just rounded them up,

seized their belongings, and forced them into a concentration camp! Anger at your government doesn't mean that you will fight for their enemy. Also, question 28 is multiple questions at once! One may have been more than happy to swear allegiance to the United States but not ready to say they would die defending it.

But nonetheless, the United States used the results of these surveys to "scientifically" evaluate the detainees. The machine was used to justify classification, incarceration, and the stripping of thousands of people's human rights and dignity.

It is worth noting that, after the war, a government investigation found that the decision to intern Japanese Americans had been due to racial prejudice, war hysteria, and a failure of political leadership.[31] There had been no credible evidence that they were a threat.[32] I can find no credible reporting that an act of domestic terror was ever plotted or carried out by Japanese Americans. There is no evidence that any of them were working as spies. By contrast, 33,000 Japanese men left their families in the camps and enlisted in the United States army.[33]

All governments are capable of using technology for evil. Often, they use it under the guise of maintaining social order and keeping the population safe. In the abstract, these missions sound virtuous, but they can easily be distorted because governments, and society at large, are often less virtuous than we like to believe.

Alan Turing was chemically castrated because he was gay, at a time when being gay was considered a threat to the social order. Japanese Americans were segregated and sent to military concentration camps because their "data" categorized them as potential threats. It is easy to see these problems as history, terrible moments that, in hindsight, we can look back on and know are wrong. But they aren't history. They are problems that are only getting worse and less transparent with the use of AI. The impacts of these problems are disproportionately felt by marginalized and racialized populations.

There are two main ways that those in power can use AI to oppress a population. The first is by intentionally using it to quash dissent. The second is through a misunderstanding of how data-driven systems such as AI operate.

It is a sometimes contradictory but well documented truth that some level of dissent is necessary in any society for social progress to occur. As Dr. Martin Luther King Jr. said, "The moral arc of the universe is long, but it bends toward justice."[34] We need the ability to push back against the immoral parts of our society, to push our boundaries and reveal where we have gone astray.

Dr. King himself was considered to be a dissident. The government ruthlessly spied on him and meddled in his life.[35] Some of his contemporaries, such as Fred Hampton, were killed by the same government that was supposed to be protecting their rights.[36] Yet, the civil rights movement ultimately prevailed.

AI provides frightening new surveillance potential. It creates the ability to monitor suspected dissidents and create data at an unprecedented level, as facial recognition and voice recognition programs can monitor subjects with no human oversight. Furthermore, using AI, surveillance agencies can try to identify potential "problem" citizens and take action against them.

This is not a hypothetical. In the United States, a program called ATLAS is used to analyze and profile US immigrants.[37] The system is designed to pull data from numerous databases, even after an immigrant is granted their status, and algorithmically determine if they pose a threat to the United States.[38] In some cases, after a person has already obtained citizenship, the system will determine that they have links to terrorism, and their citizenship will be revoked.[39]

Although a human is involved in this process, it is impossible to know the extent to which the investigators blindly trust the judgment of the system and the information it provides.[40] It is even more problematic in the case of the ATLAS program, as many of the data-bases that are used to fuel its judgments are not well maintained and have been found to contain a lot of incorrect or misleading information.[41]

This leads us to the second problem with using AI algorithms for governing populations, and honestly, for using them in any situation: data and algorithmic bias.

I am not an expert in the field of algorithmic bias. My knowledge is built upon the tireless work of brilliant scholars such as Timnit Gebru, Simone Browne, Kate Crawford, Cathy O'Neil, Elizabeth Joh, and Sarah Brayne.[42] I have done my best to do their work justice, as I believe it is an important issue for you to be aware of.

This book is centered on the human foundations of AI and our tendencies to overlook them. Nowhere is this problem more apparent than with algorithmic bias. The idea that the world can be completely stripped down to data points that can be understood and managed through statistics is a lot more hotly contested than it is often presented as being. Even if we accept that this idea is true, our current technological capabilities are FAR from being able to quantify our world. We currently struggle immensely with even modeling the flow of water accurately with statistics.[43] Trying to collect the data points and design systems that can accurately understand humans, agents with free will, is well outside our capability.

It is a common statement that data is objective. But this is plainly false. Although data in the singular sense, a single piece of data,[†] may be an objective representation of a frozen moment, data is not objective. Data is the product of human choices on what information to collect as data and what data to combine to create a dataset.[44] These choices leave human fingerprints all over data. The best example of how this creates problems is through predictive policing technology.

[†] Fun fact, a singular piece of data is called a datum.

Predictive policing, and the use of AI in policing, is all the rage right now. Police forces around the world are purchasing and using programs that claim to help them be more efficient.[45] Using AI models that claim to understand determinants of crime, and data provided by the police force using the system, predictive policing directs the police on where to best focus their resources.[46]

Some systems produce hot maps, telling cops where a crime is most likely to take place at any given moment.[47] Logically, the police divert more attention to these areas. They can also produce heat lists, providing cops with the names of individuals most likely to be involved in a violent crime so that police can keep an eye on them.[48]

The system is objective, free of human judgment. It uses advanced analytics to analyze data and direct resources. The human is taken out of the equation. Except they aren't.

The program was trained on policing data. It learned about crime, and where it is likely to occur, based on data produced by human cops. The data it learned from transfers the biases of the humans who made it into the machine.[49]

Here's a thought experiment. Imagine a small community surrounded by trailer parks. The community itself is quite wealthy, and residents live in large houses with fenced-in yards. The trailer park is impoverished, and there is a lot less privacy. Both areas have equal levels of crime. However, the crime in the wealthy area isn't as visible.

Domestic incidents occur behind closed doors. There is a lot of hard drug usage, but it happens at parties in mansions. Even when the cops are aware of illegal activity, it is more difficult to act on it because they know that the wealthy residents have the means to hire high-priced lawyers. You don't just kick down the door of a mansion with no reason. Sometimes, incidents do occur in the open. But the streets of these neighborhoods are considered to be safe by the police, so they hardly patrol the area.

On the other hand, the trailer parks are constantly patrolled. The police know and have recorded the names of most of the residents. Drug busts are common, and the police are able to identify and record a lot of crime.

Again, in this hypothetical, actual crime rates in these two communities are the same. However, social and structural forces have created a dataset where the majority of crime occurs in the trailer parks.

When an AI system is adopted, it analyzes the data and obviously comes to the conclusion that there are more drugs and violent crime in the trailer parks. It is able to use the data it has to make predictions about where the drugs are being sold and where they are being produced. This information, when acted on by the police, leads to more arrests and more intensified policing in the trailer parks. This new data is then fed back into the AI system, strengthening its assumptions.

It is fascinating because the more the trailer parks are over-policed, the more criminal actors get arrested, bringing down the trailer parks' real crime rate. Quickly, the real crime rate in the trailer parks actually becomes a fraction of that within the wealthy neighborhood. Yet the data shows the opposite. Crime is skyrocketing in the trailer parks and is at an all-time low in the mansions.

The algorithm is not objective. It just codified and justified the biases that already existed.[50]

This problem is seen across the spectrum of AI tech. It's seen in facial recognition technology misclassifying African American women as male or in an Amazon AI that processed job applications and discriminated against female candidates.[51] The Amazon system favored male candidates because it was trained with data on the sorts of people human hiring committees typically hired. Again, the bias snuck in.

There is also bias produced from the data we deem to be important to measure. In Texas, an algorithm designed to evaluate high school teachers was given the responsibility of determining who should get bonuses and who should be recommended for potential dismissal.[52] But the performance data that was used for evaluations couldn't possibly capture the full effects teachers had on their students. One history teacher who had recently won a major award for teaching achievements had his bonuses slashed and was labeled as a bad teacher.[53] But it's okay, because the machine was "objective."

This problem is made exponentially worse by the rise of AI. In the Texas school case, the teachers were able to sue the school board on constitutional grounds, arguing that their rights were violated as they were unable to observe or scrutinize how the algorithm evaluated them. The company that made the algorithm claimed that they could not demonstrate how it worked as it was a "trade secret."[54] It is likely that the algorithm in this case was not a deep learning system, and if it had been publicly "opened up," the teachers and their lawyers could have identified flaws in the system.[55] But if it were a deep learning system, there would be no way to determine exactly how the teachers were being evaluated.

Deep learning systems are not told what to look for. They are given data and an objective. The neural network structure is impossible to reverse engineer. It's just too complicated.

Reverse engineering a traditional machine learning system is like trying to navigate a corn maze. It's hard but very doable.

Reverse engineering a deep learning system is like trying to navigate a maze made of water that's the size of the Atlantic Ocean. You know where the beginning is, and you know where the end is, but there are just so many potential paths. And even though the walls of water are clear, they keep changing.

This issue is known as the black box problem.[56] There is currently no promising path to solving it. When an AI does something unexpected, there is no definitive way to know why it happened.[57] We can make informed guesses, and we can try to change our dataset, but it is an inexact science. And often, when an AI is being discriminatory, it's difficult to prove it. All we get is an output, and it can take a lot of discriminatory outputs to notice a pattern.

Part of the reason this issue is so troubling is because AI represents a new kind of tool that humans don't have much practice using.

Like the blind man with his cane, technology often becomes an extension of us. We commonly outsource tasks that are typically done by humans to machines.[58] The blind man's cane takes the job of detecting objects and relays information back to its holder. The invention of writing allowed us to take the task of remembering and passing down events and outsource it to the written word. The pen or keyboard acts as our voice, as a record of our thoughts, which will now be remembered as long as the medium we have recorded them on survives. When I was in primary school, we were forced to learn "mental math" because we wouldn't always have a calculator. But now I do always have a calculator on me, and I trust the mental job of long division or calculating a restaurant tip to it.

But the outsourcing we are doing with AI is different. We are outsourcing our ability to think, and we are outsourcing the subjective. When I write down a memory or a story, it is recorded as I wrote it. The memory is recorded directly from my mind to the paper. When I use a calculator, it is programmed to follow an objective set of rules.

I think that this previous experience has lulled us into a false sense of security, the thought that technology will do what it's supposed to do. A calculator always gives us the right answer. So it makes sense that a state-of-the-art AI will do the same. But the questions we are asking of AI are far more complex. That is why we need AI to help us with them. We need to understand that these outputs are not divine. They can be wrong, just like a human can be wrong.

ChatGPT may be able to answer philosophical questions and help build business plans, but it is a product of its data and architecture. We shouldn't outsource all of our thinking to it. I'm not saying machines shouldn't help us with tough or complex decisions. What I'm saying is that many of these jobs should be done in union by both the computer system and the person using it. As we start to rely on machines to do more thinking tasks for us, we really need to think about which tasks should be done by machines and which ones we should keep doing ourselves. We can use algorithms without letting them do all the thinking for us.

Of course, the automation of our ability to think also brings up another major problem with AI that I have thus far avoided: the automation of human work.

CHAPTER 23

Addressing the Elephant in the Room

When it comes to the future of work, there is only one certainty. Anyone who tells you that they know what is going to happen is either lying to you or trying to sell you something. We are in uncharted territory. And there is a lot of uncertainty. What we do know is that a lot is going to change.

When I first started working on AI, there were three major camps when it came to thinking about jobs:

1. AI and AI automation will progress to a point at which human workers become economically worthless.
2. We will experience a massive transition, and work as we know it will change forever. There is still a future place for humans in the workforce, and if we are smart, we can avoid the majority of human suffering associated with the transition.
3. Nothing will happen, life will go on as usual, and there will be no mass unemployment.

Before ChatGPT, most people didn't actually know much about how quickly deep learning was progressing. So position three was a pretty popular stance, and those who believed in option one were considered insane. But things have changed a lot in the past few years.

Position three is now rarely discussed. It is just assumed that AI will have a massive impact on the labor market. And although position one is still quite taboo in academia, it is generally accepted as a possibility. But most researchers, myself included, fall into the category of position two.

AI, in combination with advanced robotics research, has put a lot of jobs under the pressure of potential automation. Already, AI is being used in manufacturing, health care, therapy, management, programming, the arts, and construction, to name a few – it has basically penetrated every industry. In 2013, a now infamous Oxford University study predicted that approximately 47 percent of American jobs would be at risk of automation in the near future.[1] The researchers may have been a little ambitious, but right now it's not feeling like they were that far off.

The reality is this: A lot of tasks that were traditionally considered to be safe from automation are no longer safe.[2] Jobs that require artistic abilities or cognitive skills are now being automated. At the same time, the ability of machines to perform physical tasks is increasing as robots are fitted with AI systems that allow them to better sense and react to the world around them.[3] AI's automating powers are indiscriminate. They are affecting blue collar manufacturing jobs and white collar office jobs. Many who spent years, and thousands of dollars, developing specialized skills now need to live with the fact that AI can do their job faster and often better. It is a terrifying reality. And what makes it even scarier is that we have no clue exactly how it is going to play out. There are just too many variables.

For example, it is possible that a lot of companies will put off automating their staff and instead opt for retraining their current employees for new positions or finding ways to have them work with AI. This approach could spread out the potential economic disaster that would occur if there was suddenly mass unemployment due to rapid automation. Mass unemployment is bad (obviously) for a number of reasons. Economically, it presents a gigantic threat. If companies are producing more goods than ever due to automation, who's going to buy them? There's mass unemployment, and the people don't have money to support businesses.

Whether or not companies take this approach is not really in our hands. Even if they do, it's possible that an unrelated financial recession will occur, forcing companies to lay off their employees as is normal during a recession. However, this time it may be different as they offset these layoffs by bringing in automation technology. Now, when the recession ends, the companies have no incentive to rehire their old employees.[4]

The prefabricated argument often used to confront fears of automation is that innovation inevitably creates new jobs.[5] Although some jobs may be eliminated, the economic benefits of automation will create incentives for companies to create new jobs elsewhere in the economy; therefore, there is not actually anything to worry about!

As much as I would love to believe in this take, as it is so comforting, it unfortunately does not hold up when put under the microscope. There are several major problems that emerge under scrutiny, even if we assume that this statement is partially true and that new jobs will materialize that we cannot foresee right now. The first problems are to do with speed and scale.

When we typically talk about automation or innovation, it is in a very narrow sense. There may be a new innovation that automates a specific job in the dairy industry or in stock trading. But these instances are isolated. The automation in the dairy industry is not directly transferable to other industries. An automatic cow-milking system doesn't have disruptive powers in other industries. However, the cost savings produced by the innovation could affect other parts of the industry. As production rises, the need for new jobs in other areas of the dairy industry may be created, offsetting the lost jobs milking cows.[6]

But AI is not a one-dimensional technology. As we have discussed at length, it isn't even a single piece of technology. It is a collection of technologies and methods for collecting and processing data that can be applied in numerous ways. It is not contained to a sole application. It is what's known as a general purpose technology.[7] Other examples of general purpose technologies are the internet, computers, and most notably, the steam engine.

General purpose technologies cut across industries.[8] They do not have a sole purpose but instead can be applied to an infinite number of tasks. They become productive forces that humans interact with, and leverage, to create and innovate.

As it is a general purpose technology, AI has the ability to disrupt countless industries at once.[9] Where efficiencies in one area may have at one time resulted in new jobs in another, it is now possible that the other area will be experiencing the same efficiencies due to AI. Therefore, they may be creating some new jobs, but they will also be eliminating others. Although new jobs will be created to deal with AI, there is no certainty that they will replace all the jobs that are lost from the potential coming wave of automation.[10] But scale is not the only problem with AI's automation effects. Speed is also a major fear.[11]

AI developed in the shadows for a substantial period. Now that it has arrived in the mainstream, it seems to have entered with significant momentum, which means that it may hit countless segments of the economy fast and hard. Lots of people across industries may lose their jobs in a very short period of time. The creation of new jobs for the now unemployed workers will not happen as fast as the workers lose their jobs. In other words, if AI automation hits the job market hard and fast, we need to be prepared for a prolonged period between the layoffs and the emergence of new jobs. This period is likely to be both economically and mentally scarring for those who get caught in the middle.[12] Sadly, this situation is not where the problems end.

Again, even if we accept that new jobs will appear, and even if a miracle occurs and they appear jaw-droppingly fast, there is still a problem. The idea that new jobs will emerge brushes over the question of whether the new jobs will be appropriate for those who have lost their jobs. Not all jobs are the same.

Due to the nature of AI, it is likely that it will heavily affect workers who are deemed to be "skilled workers."[13] This does not only mean people with a university education. It also includes those who perform tasks that traditionally require a large amount of acquired knowledge that is not common.[14] This knowledge can be acquired through education or on-site experience. Workers are skilled because they have invested in themselves, acquiring this specialized knowledge. But now AI is able to acquire and act on the same knowledge. For many, massive investments they have made into acquiring knowledge and skills may be wiped out overnight. That doesn't just mean financial investments in formal education but also years of practical work and experience that previously led to them being deemed skilled or valuable in the job market. The skills that will be needed for many of the new jobs that are created in a post-AI automation world will require fundamentally different skills to those which multiple generations have trained for.[15]

The skills gap that may be created could produce gigantic economic and social problems.[16] Those who have their skills completely devalued by the economy will either need to retool or take low-skilled jobs.[17] Low-skilled jobs will not be able to provide a similar standard of living or social status as previous jobs. This downgrading will almost certainly lead to adverse mental effects and resentment of the system that betrayed these workers. That is doubly true for those who feel as if they did everything right. They did everything they were told to do to gain financial security, only to have it ripped out from under their feet. Imagine being forty-five years old with a young family and twenty-two years of work experience. Then suddenly, you are told you need to start from the bottom. Your skills are no longer valuable. The further along you are in your career, the more devastating needing to retrain can be.

However, it is not just those who are far along in their careers that are at risk. University education has long been positioned as the fastest and most efficient way to gain knowledge and skills that can set you apart. In countries such as the United States and Canada, students often take on substantial debt to fund their education. The rise of AI means that many skills taught at universities, especially in professional programs such as business, are being automated.[18] This means that recent graduates may be saddled with massive amounts of debt for training that is now practically useless.

For many, the option of retooling without specialized aid from the government will be impossible.[19] Since they still need to eat, they will be forced to take a low-skilled job. Maybe they will be supervising a team of packing robots at an Amazon warehouse or cleaning rusty gears. Whatever they end up doing, the economic numbers will show them as being employed. But that is not the full story.

Economic problems are often discussed in sterile terms. We talk about unemployment numbers and labor productivity. But it is so important to remember that behind each number is a person. That those numbers have souls, and that they love and are loved. How we as people experience the phenomenon of AI automation will affect us. It will shape our views and our values. The effects of automation can traumatize, demoralize, and even radicalize. And it is not just those who are economically displaced that are affected. Wealth inequality and economic despair affects us all. We do not live in an isolated society. What happens to your neighbors affects the world you live in too.

Moreover, financial struggles have massive effects on families. I was eleven years old when the great recession affected my family. For their privacy, I will shy away from the specific details. Yet I will be clear: I was deeply changed by what I experienced during that time. Even as my parents did their best to shield me, their youngest child, from what was happening, the effects were all too real.

To be honest, that is one of the reasons I started researching AI. Although my work eventually strayed from examining the effects of AI on jobs, it is a topic that has always made me nervous. A storm is coming, and we need to prepare. We need to ask ourselves hard questions about what we want this transition to look like and make difficult choices about how we achieve it. And we need to do it fast.

THE ROLE OF HUMANS IN THE GREAT TRANSITION

If there is one thing I hope you take from this book, one lesson that you learn, I hope it is this: Even in a world dominated by machines, humans, and our decisions, are at the center. Just as we build the machines and make decisions about how they are built, we build the systems that they are introduced into. We are not helpless bystanders when it comes to AI automation. We get to decide what the future of work looks like. We get to decide how we handle the great transition.

A group who once understood this idea was the Luddites, whom we met much earlier in the book.[20] They were skilled workers who became famous for destroying the new factory machinery that was automating their work. But as I mentioned, the story of the Luddites is much more complicated than it is often made out to be.

The Luddites were not anti-technology.[21] In fact, many Luddites embraced the new technologies and were thrilled to work alongside them. What the Luddites feared was that their bosses would not use this technology to improve the products their customers received but to cheapen them; and at the same time, they'd gain more control over their workers.

In the labor market, skilled workers have a lot of power. They can speak up for their rights and demand better conditions because they are harder to replace. By contrast, unskilled workers are easily replaced or controlled. If the company's desire is just to make money, unskilled workers are typically their preference.

The factories that employed the Luddites could have functioned, and continued to make profits, without completely replacing their workers. Although they may not have needed every skilled worker they had had before the automation, there were still opportunities to use the technology and pursue projects that required the skilled workers.

At the end of the day, the battle the Luddites fought was one over economic and personal power. They were living through a similar transition to the one we are living through today, and they didn't like the way it was being handled. They didn't like the way technology was being used to cheapen production or the way it was devaluing those who worked with it and robbing them of the potential to take pride in their work by subjecting them to an existence as just another cog in a machine – a cog that could be thrown away by the powers that be if necessary. And they especially didn't like the fact that the way these machines were being implemented was vesting massive amounts of social, economic, and political power into the hands of those who owned them.

The Luddites were living through one of the last great societal transformations, the Industrial Revolution.

There is no definitive start or end date for the Industrial Revolution. But it marks the period of human history when we completely transformed the ways we made things. Before the Industrial Revolution, we mostly made stuff by hand. Tools were used, but they were simple and needed to be wielded by someone who understood them. A large majority of the population worked as farmers, where they used their physical brawn and knowledge of the land to till the fields and feed the community.

But then the Industrial Revolution began.[22] New technologies such as the steam engine and water wheels made it possible to mechanize production. But the way production was mechanized was not set in stone. It was driven by human ideologies about how work should be done and what kind of work would lead to a better world.

Ideas about cutting costs and increasing production led to increased de-skilling of work. As more machines were introduced, humans began to be treated more like machines. Jobs were divided into small repetitive tasks. Instead of constructing a whole shoe, a worker would spend their day making the same boot part hundreds of times, every day.

The Industrial Revolution can be remembered through two competing images. On one side, it was the beginning of a gilded age. The machines of industry harnessed the productive force of society, and we began producing more goods than ever. We unlocked the potential of man and machine, and built unthinkable wealth. It gave us the ability to provide more riches for everyone.

But it was also a time of great suffering, when thousands were forced off their farms or pushed out of professions in which they had found deep meaning. Communities were shattered, and rural populations were forced into the slums of the cities to find work. With no skills, workers had little power and worked for pennies. Those who owned the factories hoarded vast wealth and understood the power their positions gave them. Corruption was rampant, and they worked hard to maintain their place at the top of the hierarchy.

Children worked in factories and starved in the streets. But they were employed. Most members of society had jobs, but were they really living? It wasn't uncommon to work ten to sixteen hours a day, six days a week, just to make ends meet. But it didn't need to be this way.

We as a society made choices about how the technology of the Industrial Revolution was to be used. We could have chosen a path that valued the worker and aimed to reduce suffering and distribute the benefits of increased production more evenly. Instead, we often chose efficiency and profit over human dignity and community. The story of the Industrial Revolution, like the story of the Luddites, is a reminder of the power of human agency in the face of technological change. It's a cautionary tale that highlights the importance of making conscious choices about how we integrate new technologies into our society, ensuring that they serve to enhance human well-being rather than diminish it.

As we stand on the brink of another transformative period, driven by AI, we are faced with similar choices. But this time, we have the past to learn from. We can see how the Industrial Revolution played out and ask, "Do we want to do that again?" Will we repeat the mistakes of the past, allowing technological progress to exacerbate inequality, displace workers without offering viable alternatives, and concentrate power and wealth in the hands of a few? Or will we learn from history and steer this technological revolution in a direction that benefits all of society?

The great transition we are undertaking is not just about the role of AI in the economy. It is not just about our jobs. It's about who we decide we want to be. It's about how we, as a society, choose to respond to the challenges and opportunities AI presents. It's about the kind of world we want to live in and the values we want to guide us. The future of work is not predetermined. How we will live with AI is not etched in stone. It is up to us to decide how we navigate this uncharted territory, ensuring that the journey leads to a destination where technology serves humanity, and not the other way around.

I have hope that AI will be a force for good, because I believe in the good in people. And at its core, AI is a very human invention.

CHAPTER 24

Hope

"I wish it need not have happened in my time," said Frodo.

"So do I," said Gandalf, "and so do all who live to see such times. But that is not for them to decide. All we have to decide is what to do with the time that is given us."

–J.R.R. Tolkien, *The Fellowship of the Ring*

We live in such times. The world is changing. We cannot fight it. It is happening whether you like it or not and whether we are ready or not. I imagine that, since you are reading this book, you feel it too.

I would be lying if I did not confess that I am often scared. I do see things that make me feel uneasy. I have expressed many fears in this book, while I have left others out. But my overwhelming emotion is not one of fear. It is one of hope. Hope that we will navigate through these transitions. Hope that we may usher in a better and more just world. I need to believe, and I choose to believe.

We, the people, get to decide what a future living with AI looks like. And those of us here right now – we are living through a historic moment. We stand at the beginning of something bigger than ourselves. And we have been given the opportunity to shape the future.

What we do now will not predefine the course of history. But it will define its borders. We get to make the rules that form the foundation of a society that coexists with AI. The choices we make now about what technologies to build and how to implement them will have massive ramifications for generations to come.

No matter how great our intentions are, we will inevitably falter. We will make mistakes that will need to be corrected along the way. We may also produce errors that compound upon themselves throughout the ages – creating untold hardship. But that is the destiny of humanity. We are perfectly imperfect. We should not let the fear of our own mistakes hold us back from dreaming of a better tomorrow.

I believe we are lucky because we are living in a moment of human history when it is not just possible to change the world but it is also necessary. Throughout history, there have been people who have called for revolutions. Some looked to violence, while others took to the streets. Those who want to see the world change often need to demand it forcefully. The status quo is stiff and unmoving. Massive social movements are needed to break the veil of normal life and usher in the political and social action necessary for change to occur.

But we need not ask for a revolution; it is already here. The question we must ask is, "What do we do about it?" What world do we want to create?

The story of AI is a human story, like the story of any technology. It is a myth that we create technology as a neutral object, which then shapes our world. We make decisions. We build technology around philosophies and ideas of the world we want to see. And we build it within the economic and legal systems built by our governments as guidelines.

I strongly believe that AI has the greatest transformational potential since the steam engine. It can be leveraged to free humans from tedious tasks. It can make us more efficient at difficult tasks. And it can usher in an era of productivity that allows every member of society to live a fulfilled life. The magic of automation is that it can free us from our own limitations – if we implement it properly.

This book is a call to action, a call to insert yourself back into the story and to see your place in the narrative of AI. It is a story that we are still writing. I do not want to tell you how to think about AI; perhaps I will write another book explicitly outlining my thoughts and aspirations for an AI world. For now, I want you to understand what AI is, how it works, what it can and cannot do, and how it is already affecting you. I want you to understand how it is reshaping our society.

I want you to understand your place in the story and where you fit into it. Ask how this technology is being implemented in your life, and then ask how you would like to see it used.

As I alluded to in the prologue, the biggest problem I see with AI is not the technology. It is the way we are deciding how it should be used. It is that a very narrow vision of what a world with AI will look like is being communicated to the public. Right now, it is those who already have power, and those who are set to benefit from the status quo, who are driving the conversation. We are told that AI is complicated, and dangerous, and that we need to trust them. But this is a conversation that belongs to everyone. It's a conversation that belongs to you.

Speak up, and speak out.

The most radical thing that you can do is become a part of the conversation. You do not need to take my position nor be concerned about the same problems I am. The issues I have highlighted in this book are the product of my interactions with the world, the products of my experiences. Your views will inherently be different, and that is good. This conversation needs complexity. We need to take into account how the decisions we are making will affect all members of our society.

I have hope because I believe in the good in people. I have hope because I believe in democracy. I believe that, if we are given the opportunity to understand where we are and how we can make a difference, we, the people, can come together and make a better world for all. I have hope because we are at the beginning of this transformation, and AI has so much potential for good.

The story of AI is not a story about machines. It is a story about people who faced problems and made decisions – decisions that, over the centuries, have accumulated to lead us to our pivotal moment.

Now we get to make the decisions. Now *we* get to shape the future and decide what kind of society we want to have.

A Note from the Author

I am not a historian, and this book is not a history. Yet that does not make it any less true. The events I discuss did happen, and they happened in the order I presented them. However, much of the complex history of AI and its development was left on the cutting room floor. Any discussion of Bell Labs, one of the most important spaces in the history of AI development, was confined to a small footnote. We did not discuss the early AI research at Xerox PARC, competing historical theories regarding the invention of the algorithm, or the work of Soviet scientists and their collaborations with the founding fathers of AI. Further, in many ways I have painted the symbolic AI researchers as a caricature of who they were and failed to meaningfully discuss many of their triumphs from the Logic Theorist to Block World. Although I discuss Hinton in depth, I ignore many of his contemporaries who are just as important to the development of AI as he and his team. For each omission, I have my justification; yet a detailed writing of each justification would likely be long enough to publish as a trilogy of books.

I am confident that there will be countless critiques of stories that I left out or smoothed over. Yet I am content with the story I have written.

I left these stories out because, although many may read this book as a history of AI, it is not. It is a history of the present. I did not write this book to inform you about all of the technical advancements or people that led to our current moment. That was not my goal. My goal was to give you the knowledge about AI that you need to participate in democracy today – the knowledge you need to understand the current effects of AI on society, the current effects of AI on your life. You do not need all of the historical details to understand AI as it is. I used history as a ship to navigate complicated and rough waters. The stories I chose to highlight were chosen as they best allowed me to illustrate how AI has developed to this point and to pull back the curtain on its mythicization. I centered Hinton's story not because he is the most important or the reason for our current moment, but because he provided a vessel for me to discuss and present the grander societal narrative of which AI is a part. I sincerely hope that reading this book has allowed you to develop a greater understanding of AI and its social effects. If you are interested in the complete history of AI, there is no reason your journey must end here.

Notes

1. The Household Name That No One Knows

1 D.M. Dunlop, "Muḥammad b. Mūsā al-Khwārizmī," *Journal of the Royal Asiatic Society of Great Britain and Ireland*, no. 2 (1943): 250, https://www.jstor.org/stable/25221920.

2 A.B. Arndt, "Al-Khwarizmi," *The Mathematics Teacher* 76, no. 9 (1983): 668–9, https://www.jstor.org/stable/27963784.

3 Roshdi Rashed, *Encyclopedia of the History of Arabic Science* (Routledge, 1996), 2:17, https://archive.org/details/RoshdiRasheded.EncyclopediaOfTheHistoryOfArabicScienceVol.3Routledge1996.

4 Jason Goodwin, "The Glory That Was Baghdad," *The Wilson Quarterly* 27, no. 2 (2003): 24, https://www.jstor.org/stable/40261181.

5 Matthew E. Falagas, Effie A. Zarkadoulia, and George Samonis, "Arab Science in the Golden Age (750–1258 C.E.) and Today," *The FASEB Journal* 20, no. 10 (August 2006): 1581–6, https://doi.org/10.1096/fj.06-0803ufm.

6 Goodwin, "The Glory That Was Baghdad," 24.

7 Philip Maher, "From Al-Jabr to Algebra," *Mathematics in School* 27, no. 4 (1998): 14–15, https://www.jstor.org/stable/30211868; Rashed, *Encyclopedia of the History of Arabic Science*, 2:17.

8 Arndt, "Al-Khwarizmi," 670; Rashed, *Encyclopedia of the History of Arabic Science*, 2:3.

9 Arndt, 670; Rashed, 2:3.

10 C.E. Bosworth, "A Pioneer Arabic Encyclopedia of the Sciences: Al Khwārizmī's Keys of the Sciences," *Isis* 54, no. 1 (1963): 99, https://www.jstor.org/stable/228730.

11 David S. Powers, "The Islamic Inheritance System: A Socio-Historical Approach," *Arab Law Quarterly* 8, no. 1 (1993): 13–29, https://doi.org/10.2307/3381490.

12 Rashed, *Encyclopedia of the History of Arabic Science*, 2:18.

13 Jeffrey A. Oaks, "Medieval Arabic Algebra as an Artificial Language," *Journal of Indian Philosophy* 35, no. 5/6 (2007): 546, https://www.jstor.org/stable/23497285;

"In the preamble of his book, al-Khwārizmī mentions the generous encouragement of the arts and sciences by the Caliph al-Ma'mūn, who had encouraged him to write his book" (Rashed, *Encyclopedia of the History of Arabic Science*, 2:37).

14 Muḥammad ibn Mūsā al-Khuwārizmī, *The Compendious Book on Calculation by Completion and Balancing* (Paul Barber Press, 1937), retrieved from the Library of Congress, https://lccn.loc.gov/2021666184.

15 Arndt, "Al-Khwarizmi," 670.

16 In the introduction to his book, al-Khwārizmī directly states that he "wrote from the work on algebra a brief book which encompasses the fine and important parts of its calculations that people constantly require in cases of their inheritance, their legacies ... and in all their dealings with one another." Quoted in Oaks, "Medieval Arabic Algebra," 545–6.

17 Lam Lay Yong, "The Development of Hindu-Arabic and Traditional Chinese Arithmetic," *Chinese Science*, no. 13 (1996): 36, https://www.jstor.org/stable /43290379; Arndt, "Al-Khwarizmi," 670.

2. The Enchantress of Numbers

1 Thomas M. Disch, "My Roommate Lord Byron," *The Hudson Review* 54, no. 4 (2002): 590, https://doi.org/10.2307/3853312.

2 Lady Caroline Lamb, quoted by Lady Sydney Morgan in her memoir: Lady Sydney Morgan, *Lady Morgan's Memoirs: Autobiography, Diaries and Correspondence* (W.H. Allen, 1863), 2:200.

3 Betty Alexandra Toole, *Ada, the Enchantress of Numbers: Poetical Science* (Critical Connection, 2010), 40, Apple Books; William Edwar West and Estill Curtis Pennington, "Painting Lord Byron: An Account by William Edward West," *Archives of American Art Journal* 24, no. 2 (1984): 16–21, https://www.jstor.org/stable /1557222.

4 Brian Merchant, *Blood in the Machine: The Origins of the Rebellion against Big Tech* (Little, Brown, 2023), 585–8, Apple Books.

5 Merchant, *Blood in the Machine*, 585–8.

6 Mary Shelly, "Introduction," in *Frankenstein* (MIT Press, 1831), https://doi.org /10.7551/mitpress/10815.001.0001.

7 Toole, *Ada, the Enchantress of Numbers*, 54.

8 Toole, 54.

9 Toole, 42.

10 B.H. Neumann, "Byron's Daughter," *The Mathematical Gazette* 57, no. 400 (June 1973): 94, https://doi.org/10.2307/361534. For further reading, see Toole, *Ada, the Enchantress of Numbers*, chapters 1–2; Ethel Colburn Mayne, *The Life and Letters of Anne Isabella, Lady Noel Byron*, 2nd ed. (Constable, 1929), http://archive.org /details/in.ernet.dli.2015.184379.

11 Neumann, "Byron's Daughter," 94.

12 In an affectionate nod to Annabella's mathematical prowess, early in their courtship Byron referred to her as his "Princess of Parallelograms." Once the relationship soured, however, so too did Byron's characterization of her. Instead of a "Princess," Byron described Annabella as a calculating "Mathematical Medea"; he later elaborated upon this bitter view in his famous poem *Don Juan*. See Catherine M. Andronik, *Wildly Romantic: The English Romantic Poets – The Mad, the Bad, and the Dangerous*

(Henry Holt, 2007), http://archive.org/details/wildlyromanticen0000andr; Maria Popova, "How Ada Lovelace, Lord Byron's Daughter, Became the World's First Computer Programmer," *The Marginalian*, December 10, 2014, https://www .themarginalian.org/2014/12/10/ada-lovelace-walter-isaacson-innovators/; Toole, *Ada, the Enchantress of Numbers*, chapter 1.

13 Toole, *Ada, the Enchantress of Numbers*, 37.

14 Toole, 40.

15 Toole, 42.

16 Toole, 37.

17 Merchant, *Blood in the Machine*, 38.

18 Merchant, 698.

19 Toole, *Ada, the Enchantress of Numbers*, 54.

20 Herman H. Goldstine, "A Brief History of the Computer," *Proceedings of the American Philosophical Society* 121, no. 5 (1977): 341, https://www.jstor.org/stable/986336; Merchant, *Blood in the Machine*, 698.

21 Toole, *Ada, the Enchantress of Numbers*, 43.

22 Neumann, "Byron's Daughter."

23 Toole, *Ada, the Enchantress of Numbers*, 75.

24 Martin Campbell-Kelly et al., *Computer: A History of the Information Machine*, 3rd ed. (Routledge, 2014), 28, Apple Books, https://www.taylorfrancis.com/books /mono/10.4324/9780429495373/computer-martin-campbell-kelly-william -aspray-jeffrey-yost-nathan-ensmenger.

25 Charles Babbage, *Passages from the Life of a Philosopher* (Longman, Green, Longman, Roberts, & Green, 1864), chapter 5, 2.

26 Babbage, *Passages*, chapter 5, 2

27 Campbell-Kelly et al., *Computer*, chapters 1 and 3.

28 Campbell-Kelly et al., 100.

29 Kevin C. Knox and Richard Noakes, *From Newton to Hawking: A History of Cambridge University's Lucasian Professors of Mathematics* (Cambridge University Press, 2003).

30 The original government grant, received in 1823, allocated £1,500 to Babbage's project, but by 1833, the government had spent over £17,000, "and Babbage claimed to have spent much the same again from his own pocket." Funding began to dry up, and in 1834 Babbage requested more money from the government. Unfortunately, the government felt that Babbage's results had been lackluster in proportion to the funds they'd already invested. Perhaps more detrimental to his request, in his letter Babbage noted his interest in an entirely different engine, the Analytical Engine. This was the final nail in the coffin; the government lost confidence in Babbage's focus, and funding was cut for good. Campbell-Kelly et al., *Computer*, 33–5.

31 Campbell-Kelly et al., 33.

32 Campbell-Kelly et al., 102.

33 Robert Scoble, *A Demo of Charles Babbage's Difference Engine*, YouTube, 2010, https://www.youtube.com/watch?v=BlbQsKpq3Ak.

34 Campbell-Kelly et al., *Computer*, 90.

35 Neumann, "Byron's Daughter," 94–7.

36 Eugene Eric Kim and Betty Alexandra Toole, "Ada and the First Computer," *Scientific American* 280, no. 5 (1999): 78, https://www.jstor.org/stable/26058246.

37 Sophia Elizabeth De Morgan, *Memoir of Augustus De Morgan* (Longmans, Green, 1882), 89, http://archive.org/details/memoirofaugustus00demorich.

38 Stephen Wolfram, "Untangling the Tale of Ada Lovelace," *Stephen Wolfram Writings*, December 10, 2015, https://writings.stephenwolfram.com/2015/12/untangling-the-tale-of-ada-lovelace/.

39 Neumann, "Byron's Daughter," 94.

40 Toole, *Ada, the Enchantress of Numbers*, 132.

41 Christopher Hollings, Ursula Martin, and Adrian Rice, "Ada Lovelace and the Analytical Engine," *Ada Lovelace: Bodleian Libraries, University of Oxford* (blog), July 26, 2018, https://blogs.bodleian.ox.ac.uk/adalovelace/2018/07/26/ada-lovelace-and-the-analytical-engine/.

42 Campbell-Kelly et al., *Computer*, 104.

43 Kim and Toole, "Ada and the First Computer," 78.

44 Luigi Federico Menabrea, "Sketch of the Analytical Engine Invented by Charles Babbage: With Notes by the Translator," in *Scientific Memoirs*, trans. Ada Augusta, Countess of Lovelace (Richard and John E. Taylor, 1843), 3:694, https://doi.org/2809523.2809528.

45 "Ada saw something that Babbage in some sense failed to see. In Babbage's world his engines were bound by number ... What Lovelace saw ... is the fundamental transition from calculation to computation – to general-purpose computation." Doron Swade, quoted in Matteo Pasquinelli, *The Eye of the Master: A Social History of Artificial Intelligence* (Verso, 2023), 156–7, Apple Books.

46 It is said that "the first 'digital picture' – that is, an image described by a numerical file – [was] an 1839 portrait of Jacquard himself that was woven using ... these punched cards. Babbage kept a copy of Jacquard's portrait in his own studio" and used the same technology in his Analytical Engine thereafter. Quoted in Pasquinelli, *The Eye of the Master*, 122; Kim and Toole, "Ada and the First Computer," 78.

47 Toole, *Ada, the Enchantress of Numbers*, 83.

48 To this end, Ada mused about the looms, saying that "this Machinery reminds me of Babbage and his gem of all mechanism." Quoted in Walter Isaacson, *The Innovators: How a Group of Hackers, Geniuses, and Geeks Created the Digital Revolution* (Simon and Schuster, 2014), 14.

49 Ada Lovelace, "notes" to Menabrea, "Sketch of the Analytical Engine," 696.

50 As it relates to the story of Babbage's engines, against the backdrop of the political climate – in 1832, Babbage published his book *On the Economy of Machinery and Manufactures*. The book ultimately "argued that factories and machinery were the great drivers of prosperity." However, as contemporarily relevant as his book was, "Babbage's rhetoric never explicitly acknowledged the Machinery Question – the public debate ... Rather, for him ... the new science of automated calculation [was] meant to serve solely as a multiplier of productivity." Quotes in Merchant, *Blood in the Machine*, and Pasquinelli, *The Eye of the Master*, respectively.

51 Merchant, *Blood in the Machine*, 688–9.

52 For information on the "Luddite Fallacy," see Tom Lehman, "Countering the Modern Luddite Impulse," *The Independent Review* 20, no. 2 (2015): 265–83, https://www.independent.org/tir/2015-fall/countering-the-modern-luddite-impulse/.

53 In the end, Ada never formally "raised the problem of the substitution of weavers' intelligence by a series of automatic program cards, nor the consequent sufferings of London's skilled unemployed." However, the sociopolitical conversations that ran parallel to her work – typified by the fears and ideas of the Luddites – remain inextricable from her story, both then and today. Simon Schaffer, quoted in Pasquinelli, *The Eye of the Master*, 135.

54 Kim and Toole, "Ada and the First Computer," 78.

55 Menabrea, "Sketch of the Analytical Engine."

56 Menabrea, 722.

57 Kim and Toole, "Ada and the First Computer," 76.

58 David Leavitt, *The Man Who Knew Too Much: Alan Turing and the Invention of the Computer* (W.W. Norton, 2006), 9.

59 Leavitt, 47.

3. War Hero

1 Andrew Hodges, *Alan Turing: The Enigma* (Random House, 2012), 20, Apple Books.

2 David Leavitt, *The Man Who Knew Too Much: Alan Turing and the Invention of the Computer* (W.W. Norton, 2006), 9.

3 Maxwell Herman Alexander Newman, "Alan Mathison Turing, 1912–1954," *Biographical Memoirs of Fellows of the Royal Society* 1, no. 1 (1955): 253–63, https://doi.org/10.1098/rsbm.1955.0019, 253.

4 Hodges, *Alan Turing: The Enigma*, 151.

5 Leavitt, *The Man Who Knew Too Much*, 20.

6 Hodges, *Alan Turing: The Enigma*, 151.

7 Hodges, 125.

8 Alan Turing, "On Computable Numbers, with an Application to the Entscheidungsproblem," *Proceedings of the London Mathematical Society* 2, no. 42 (1937): 230–65, https://doi.org/10.1112/plms/s2-42.1.230.

9 For further reading, see Algis Valiunas, "Turing and the Uncomputable," *The New Atlantis*, no. 61 (2020): 44–75, https://www.jstor.org/stable/26898500; Steve Russ, "Review of *The Universal Turing Machine: A Half-Century Survey*, by Rolf Herken," *The British Journal for the History of Science* 22, no. 4 (1989): 451–2, https://www.jstor.org/stable/4026930; R.K. Shyamasundar, "The Computing Legacy of Alan M. Turing (1912–1954)," *Current Science* 106, no. 12 (2014): 1669–80, https://www.jstor.org/stable/24103000; Guido Gherardi, "Alan Turing and the Foundations of Computable Analysis," *The Bulletin of Symbolic Logic* 17, no. 3 (2011): 394–430, https://www.jstor.org/stable/41228533.

10 Newman, "Alan Mathison Turing, 1912–1954."

11 Christopher Smith, "How I Learned to Stop Worrying and Love the Bombe: Machine Research and Development and Bletchley Park," *History of Science; An Annual Review of Literature, Research and Teaching* 52, no. 2 (May 2014): 200–22, https://doi.org/10.1177/0073275314529861.

12 Hodges, *Alan Turing: The Enigma*.

13 "Artifacts – Enigma Machine," Museum: Central Intelligence Agency, accessed January 17, 2025, https://www.cia.gov/legacy/museum/artifact/enigma-machine/.

14 Smith, "How I Learned to Stop Worrying and Love the Bombe."

15 "Alan Turing's Legacy," *The Royal British Legion* (blog), February 13, 2019, https://www.britishlegion.org.uk/stories/alan-turing-s-legacy-codebreaking-computing-and-turing-s-law; "War of Secrets: Cryptology in WWII," National Museum of the United States Air Force.

16 Kim Knight, "Enigma Code: Innovation That Saved 14 Million Lives," *The Mobility Forum* (blog), June 11, 2020, https://themobilityforum.net/2020/06/11/enigma-code-innovation-that-saved-14-million-lives/. For more reading on the impact of

the Bombe on the Second World War, see Peter Calvocoressi, *Top Secret Ultra: The Full Story of Ultra and Its Impact on World War II* (Sphere Books, 1981).

17 B. Jack Copeland and Diane Proudfoot, "What Turing Did After He Invented the Universal Turing Machine," *Journal of Logic, Language, and Information* 9, no. 4 (2000): 491–509, https://www.jstor.org/stable/40180239.

4. The War

1 Andrew Hodges, *Alan Turing: The Enigma* (Random House, 2012), 102–23, Apple Books.

2 Hodges, *Alan Turing: The Enigma*, 74–123.

3 Hodges, 74–123.

4 "Christopher Morcom (1911–1930)," *The Old Shirburnian Society* (blog), February 8, 2023, https://oldshirburnian.org.uk/christopher-morcom-1911-1930/.

5 "Christopher Morcom."

6 David Leavitt, *The Man Who Knew Too Much: Alan Turing and the Invention of the Computer* (W.W. Norton, 2006), 16.

7 Hodges, *Alan Turing: The Enigma*, 124.

8 Hodges, 125.

9 Hodges, 126.

10 Leavitt, *The Man Who Knew Too Much*, 17.

11 Hodges, *Alan Turing: The Enigma*, 131.

12 Alan Turing, "Nature of Spirit," an essay written during a visit to Clock House, Bromsgrove (home of his friend Christopher Morcom), April 1932. Digital copy available from the Turing Digital Archive, https://turingarchive.kings.cam.ac.uk /unpublished-manuscripts-and-drafts-amtc/amt-c-29.

13 Hodges, *Alan Turing: The Enigma*, 157.

14 Alan Turing, "Computing Machinery and Intelligence," *MIND: A Quarterly Review of Psychology and Philosophy* 59, no. 236 (October 1950): 2–40, https://turingarchive .kings.cam.ac.uk/publications-lectures-and-talks-amtb/amt-b-9.

15 Turing, "Computing Machinery and Intelligence," 2.

16 Kevin Warwick and Huma Shah, *Turing's Imitation Game: Conversations with the Unknown* (Cambridge University Press, 2016), 28–9, https://doi.org/10.1017 /CBO9781107297234.

17 Algis Valiunas, "Turing and the Uncomputable," *The New Atlantis*, no. 61 (2020): 44–75, https://www.jstor.org/stable/26898500.

18 Lee A. Gladwin, "Alan Turing, Enigma, and the Breaking of German Machine Ciphers in World War II," *Prologue* 29, no. 3 (Fall 1997): 216, https://archivesearch .lib.cam.ac.uk/repositories/7/archival_objects/272723.

19 Warwick and Shah, *Turing's Imitation Game*, 29.

20 To this end, in his paper, Turing states his belief that "the original question, 'Can machines think!' ... [is] too meaningless to deserve discussion" (Turing, "Computing Machinery and Intelligence," 442).

21 Luigi Federico Menabrea, "Sketch of the Analytical Engine Invented by Charles Babbage: With Notes by the Translator," in *Scientific Memoirs*, trans. Ada Augusta, Countess of Lovelace (Richard and John E. Taylor, 1843), https://www.fourmilab. ch/babbage/sketch.html. Quoted in Matteo Pasquinelli, *The Eye of the Master: A Social History of Artificial Intelligence* (Verso, 2023).

22 Turing, "Computing Machinery and Intelligence," 14.

23 Bernardo Gonçalves, "Can Machines Think? The Controversy That Led to the Turing Test," *AI and Society* 38, no. 6 (2023): 2499–509, https://doi.org/10.1007/s00146-021-01318-6.

24 Samuel Gibbs, "AlphaZero AI Beats Champion Chess Program After Teaching Itself in Four Hours," *The Guardian*, December 7, 2017, https://www.theguardian.com/technology/2017/dec/07/alphazero-google-deepmind-ai-beats-champion-program-teaching-itself-to-play-four-hours.

25 Objecting to the comparison between machines and a human brain, Geoffrey Jefferson introduced the example of a machine's ability to "write a sonnet or a concerto because of thoughts and emotions felt." Turing publicly responded to this question, and a public exchange between the two ensued throughout 1949, culminating in a formal response by Turing in his 1950 paper "Computing Machinery and Intelligence." See more about this discourse in Gonçalves, "Can Machines Think?," 8–11.

26 Stuart J. Russell and Peter Norvig, *Artificial Intelligence: A Modern Approach*, 3rd ed., Prentice Hall Series in Artificial Intelligence (Pearson Education, 2010), viii (Preface).

27 Russell and Norvig, *Artificial Intelligence*, viii (Preface).

28 Bryan House, "2012: A Breakthrough Year for Deep Learning," Medium, *Deep Sparse* (blog), July 17, 2019, https://medium.com/neuralmagic/2012-a-breakthrough-year-for-deep-learning-2a31a6796e73.

5. Setting the Board

1 Recounting exactly what his father would say, Hinton said, "Every morning when I went to school he'd actually say to me, as I walked down the driveway, 'get in there pitching and maybe when you're twice as old as me you'll be half as good.'" See Scott Pelley, "Godfather of Artificial Intelligence: Geoffrey Hinton on the Promise, Risks of Advanced AI," *60 Minutes* June 16, 2024, https://www.cbsnews.com/news/geoffrey-hinton-ai-dangers-60-minutes-transcript/.

2 Joshua Rothman, "Why the Godfather of A.I. Fears What He's Built," *The New Yorker*, November 13, 2023, https://www.newyorker.com/magazine/2023/11/20/geoffrey-hinton-profile-ai.

6. The Opening

1 Thomas S. Kuhn, *The Structure of Scientific Revolutions*, 2nd ed. (University of Chicago Press, 1970).

2 Marvin Minsky, "A Neural-Analogue Calculator Based upon a Probability Model of Reinforcement," Technical Report (Harvard University Psychological Laboratories, January 1952), 7.

3 SNARC, an abbreviation for Stochastic Neuro-Analog Reinforcement Computer.

4 Jozef Kelemen, "From Artificial Neural Networks to Emotion Machines with Marvin Minsky," *Acta Polytechnica Hungarica* 4, no. 4 (2007), https://acta.uni-obuda.hu/Kelemen_12.pdf; Conrad D. James et al., "A Historical Survey of Algorithms and Hardware Architectures for Neural-Inspired and Neuromorphic Computing Applications," *Biologically Inspired Cognitive Architectures* 19 (January 1, 2017): 6, https://doi.org/10.1016/j.bica.2016.11.002.

5 Warren S. McCulloch and Walter Pitts, "A Logical Calculus of the Ideas Immanent in Nervous Activity," *Bulletin of Mathematical Biophysics* 5 (1943): 115–33, https://doi.org/10.1007/BF02478259.

6 Gualtiero Piccinini, "The First Computational Theory of Mind and Brain: A Close Look at McCulloch and Pitts's 'Logical Calculus of Ideas Immanent in Nervous Activity,'" *Synthese* 141, no. 2 (2004): 176, https://www.jstor.org/stable/20118476; Tara H. Abraham, "(Physio)logical Circuits: The Intellectual Origins of the McCulloch–Pitts Neural Networks," *Journal of the History of the Behavioral Sciences* 38, no. 1 (December 2002): 18, https://doi.org/10.1002/jhbs.1094.

7 McCulloch and Pitts, "A Logical Calculus," 115–33.

8 Abraham, "(Physio)logical Circuits," 19.

9 "The name Boolean algebra (or Boolean 'algebras') for the calculus originated by Boole, extended by Schröder, and perfected by Whitehead seems to have been first suggested by Sheffer, in 1913." Edward V. Huntington, "A New Set of Independent Postulates for the Algebra of Logic with Special Reference to Whitehead and Russell's *Principia Mathematica*," *Transactions of the American Mathematical Society* 35, no. 1 (1933): 278, https://doi.org/10.1090/S0002-9947-1933-1501684-X. For further reading on McCulloch and Pitts's use of Boolean algebra, see Abraham, "(Physio)logical Circuits," 19.

10 Pamela McCorduck, *Machines Who Think: A Personal Inquiry into the History and Prospects of Artificial Intelligence*, 2nd ed. (CRC Press, 2004).

11 Minsky, "A Neural-Analogue Calculator," 7.

12 Allen Newell and Herbert Simon, "The Logic Theory Machine – A Complex Information Processing System," *IRE Transactions on Information Theory* 2, no. 3 (September 1956): 61–79, https://doi.org/10.1109/TIT.1956.1056797.

13 McCorduck, *Machines Who Think*.

14 Newell and Simon, "The Logic Theory Machine."

15 Leo Gugerty, "Newell and Simon's Logic Theorist: Historical Background and Impact on Cognitive Modeling," *Proceedings of the Human Factors and Ergonomics Society Annual Meeting* 50, no. 9 (October 2006): 880–4, https://doi.org/10.1177/154193120605000904.

16 McCorduck. *Machines Who Think*, chapter 5.

17 McCorduck, chapter 5.

18 McCorduck, 114–15.

19 Mikel Olazaran, "A Sociological Study of the Official History of the Perceptrons Controversy," *Social Studies of Science* 26, no. 3 (1996): 617, https://www.jstor.org/stable/285702; Matteo Pasquinelli, *The Eye of the Master: A Social History of Artificial Intelligence* (Verso, 2023), 41–2, Apple Books.

20 Olazaran, "Perceptrons Controversy," 617; Pasquinelli, *The Eye of the Master*, 41–2.

7. Let the Middle Game Begin

1 Frank Rosenblatt, "The Perceptron: A Perceiving and Recognizing Automaton (Project Para)," Report no. 85-460-1, Cornell Aeronautical Laboratory, 1957, https://bpb-us-e2.wpmucdn.com/websites.umass.edu/dist/a/27637/files/2016/03/rosenblatt-1957.pdf.

2 Frank Rosenblatt, "The Perceptron: A Probabilistic Model for Information Storage and Organization in the Brain," *Psychological Review* 65, no. 6 (1958): 386–408, https://doi.org/10.1037/h0042519.

3 Frank Rosenblatt, *Principles of Neurodynamics: Perceptrons and the Theory of Brain Mechanisms*, Cornell Aeronautical Laboratory, Report no. VG-1196-G-8 (Spartan Books, 1961).

4 The first version of the Perceptron was run in 1957 on the five-ton IBM 704 computer. In this first iteration, "the computer was fed a series of punch cards … [which it] taught itself to distinguish" between. Later versions of the Perceptron, as described in this text, ran on the infamous 1960 Mark I. The architecture and configuration of the Mark I Perceptron, such as the methods for inputs and outputs, varied experimentally. For this quote and further reading, see Matteo Pasquinelli, *The Eye of the Master: A Social History of Artificial Intelligence* (Verso, 2023), chapter 9, Apple Books. See also Melanie Lefkowitz, "Professor's Perceptron Paved the Way for AI," *Cornell Chronicle*, September 25, 2019; "Machine Learns Alphabet," *Science News* 78, no. 1 (July 2, 1960): 7, https://doi.org/10.2307/3942359; Nils J. Nilsson, *The Quest for Artificial Intelligence* (Cambridge University Press, 2009), 4.2.1, https://doi.org/10.1017/CBO9780511819346.

5 Herman H. Goldstine, "Computers and Perception," *Proceedings of the American Philosophical Society* 108, no. 4 (1964): 287–8, https://www.jstor.org/stable/985908.

6 Goldstine, "Computers and Perception," 287–8.

7 Frank Rosenblatt, "The Design of an Intelligent Automaton," *Research Trends: Cornell Aeronautical Laboratory* 6, no. 2 (1958), http://archive.org/details/sim_research-trends_summer-1958_6_2.

8 Nilsson, *The Quest for Artificial Intelligence*, 97.

9 Mikel Olazaran, "A Sociological Study of the Official History of the Perceptrons Controversy," *Social Studies of Science* 26, no. 3 (1996): 611–59, https://www.jstor.org/stable/285702; Pamela McCorduck, *Machines Who Think: A Personal Inquiry into the History and Prospects of Artificial Intelligence*, 2nd ed. (CRC Press, 2004), 106.

10 Olazaran, "Perceptrons Controversy," 627; Stuart J. Russell and Peter Norvig, *Artificial Intelligence: A Modern Approach*, 3rd ed. (Pearson Education, 2010), 761.

11 Evelyn Fox Keller, "Organisms, Machines, and Thunderstorms: A History of Self-Organization, Part One," *Historical Studies in the Natural Sciences* 38, no. 1 (February 2008): 72–3, https://dspace.mit.edu/handle/1721.1/50990.

12 Optimistic Futurology, *The History of Artificial Intelligence [Documentary]*, YouTube, 2020, https://www.youtube.com/watch?v=R3YFxF0n8n8&ab_channel=Futurology%E2%80%94AnOptimisticFuture.

13 Melanie Mitchell, *Artificial Intelligence: A Guide for Thinking Humans* (Farrar, Straus and Giroux, 2019), 32.

14 Mitchell, 32–3.

15 Mitchell, 32–3; Pasquinelli, *The Eye of the Master*, 477.

16 Amirhosein Toosi et al., "A Brief History of AI: How to Prevent Another Winter (A Critical Review)," *PET Clinics* 16, no. 4 (September 6, 2021): 7–8, https://doi.org/10.1016/j.cpet.2021.07.001.

17 Toosi et al., "A Brief History of AI," 9; for further reading on Deep Blue, see Monty Newborn, *Deep Blue: An Artificial Intelligence Milestone* (Springer, 2003), https://doi.org/10.1007/978-0-387-21790-1.

18 Toosi et al., "A Brief History of AI," 13.

19 Toosi et al., 13.

20 Rosenblatt, "The Design of an Intelligent Automaton," 3.

8. The Hippie at My Door

1 George Boole, *The Mathematical Analysis of Logic: Being an Essay Towards a Calculus of Deductive Reasoning* (Macmillan, Barclay, & Macmillan, 1847), https://www.gutenberg.org/ebooks/36884.
2 Will Henshall, "TIME100 AI 2023: Geoffrey Hinton," *Time,* September 7, 2023, https://time.com/collection/time100-ai/6309026/geoffrey-hinton/.
3 Henshall, "TIME100 AI 2023: Geoffrey Hinton."
4 Richard L. Gregory and John N. Murrell, "Hugh Christopher Longuet-Higgins, 11 April 1923–27 March 2004," *Biographical Memoirs of Fellows of the Royal Society* 52 (December 1, 2006): 149–66, https://doi.org/10.1098/rsbm.2006.0012.
5 Gregory and Murrell, "Hugh Christopher Longuet-Higgins."
6 Gregory and Murrell, 153.
7 Scott Pelley, "Godfather of Artificial Intelligence: Geoffrey Hinton on the Promise, Risks of Advanced AI," *60 Minutes,* June 16, 2024, https://www.cbsnews.com/news/geoffrey-hinton-ai-dangers-60-minutes-transcript/.
8 BVVA Foundation, "Geoffrey Hinton, 9th Frontiers of Knowledge Award in Information and Communication Technologies," Frontiers of Knowledge Awards, https://www.frontiersofknowledgeawards-fbbva.es/galardonados/geoffrey-hinton-2/; Henshall, "TIME100 AI 2023: Geoffrey Hinton."
9 BBVA Foundation, "Geoffrey Hinton, 9th Frontiers of Knowledge Award"; Henshall, "TIME100 AI 2023: Geoffrey Hinton."
10 Amirhosein Toosi et al., "A Brief History of AI: How to Prevent Another Winter (A Critical Review)," *PET Clinics* 16, no. 4 (September 6, 2021), https://doi.org/10.1016/j.cpet.2021.07.001.
11 BBVA Foundation, "Geoffrey Hinton, 9th Frontiers of Knowledge Award."
12 Melanie Mitchell, *Artificial Intelligence: A Guide for Thinking Humans* (Farrar, Straus and Giroux, 2019), 24–32.
13 BBVA Foundation, "Geoffrey Hinton, 9th Frontiers of Knowledge Award."
14 Terrence J. Sejnowski, "9 Convolutional Learning," in *The Deep Learning Revolution* (MIT Press, 2018), 127–42, https://ieeexplore.ieee.org/document/8555416.
15 Mitchell, *Artificial Intelligence,* 40–2.
16 Mitchell, 39–40.
17 Mitchell, 42.

9. The Machine

1 John H. Lienhard, "Inventing the Computer," Broadcast Transcript, *The Engines of Our Ingenuity* (Houston Public Media), https://engines.egr.uh.edu/episode/1059.
2 Jack Schofield, "Ken Olsen Obituary," *The Guardian,* February 9, 2011, https://www.theguardian.com/technology/2011/feb/09/ken-olsen-obituary.
3 Martin Campbell-Kelly et al., *Computer: A History of the Information Machine,* 3rd ed. (Routledge, 2014), 117, https://doi.org/10.4324/9780429495373.
4 Thomas Haigh and Paul E. Ceruzzi, *A New History of Modern Computing* (MIT Press, 2021), 181, https://direct.mit.edu/books/monograph/5194/A-New-History-of-Modern-Computing.
5 For further reading on the integration and development of these features, see chapters 6 and 7 in Haigh and Ceruzzi, *A New History of Modern Computing.*

6 Campbell-Kelly et al., *Computer*.
7 1971 advertisement by Intel, reproduced on their website: "Announcing a New Era of Integrated Electronics: The Intel 4004," Intel, 1971, https://www.intel .com/content/www/us/en/history/virtual-vault/articles/the-intel-4004.html; Campbell-Kelly et al., *Computer*, 232.
8 Intel, "The Intel 4004"; Campbell-Kelly et al., *Computer*, 232.
9 In 1965, Douglas Engelbart developed the computer mouse while working with his research team at SRI International. In addition to the mouse, the group has made significant contributions to the computing technology we use today, including hypertext, networked computers, word processing, and precursors to graphical user interfaces. For further reading, see chapter 11 in Campbell-Kelly et al., *Computer*; and chapters 6 and 9 in Haigh and Ceruzzi, *A New History of Modern Computing*.
10 Campbell-Kelly et al., *Computer*, 465–70.
11 Haigh and Ceruzzi, *A New History of Modern Computing*, 182.
12 Campbell-Kelly et al., *Computer*, 509.
13 Haigh and Ceruzzi, *A New History of Modern Computing*, 245.
14 Haigh and Ceruzzi, 245.
15 Campbell-Kelly et al., *Computer*, 509.
16 Haigh and Ceruzzi, *A New History of Modern Computing*, 255.
17 Haigh and Ceruzzi, *A New History of Modern Computing*, 25; Campbell-Kelly et al., *Computer*, 509.
18 Walter Isaacson, *Steve Jobs* (Simon and Schuster, 2011), 193.
19 Campbell-Kelly et al., *Computer*, 509.
20 Isaacson, *Steve Jobs*, 133.
21 Haigh and Ceruzzi, *A New History of Modern Computing*, 256.
22 Haigh and Ceruzzi, 258.
23 Bill Gates, quoted during "The Dating Game" at a 1983 Apple event, https://www .youtube.com/watch?v=4qtYaRrJOXc&ab_channel=Slate.
24 Campbell-Kelly et al., *Computer*, 461–2.
25 Campbell-Kelly et al., 524.
26 Haigh and Ceruzzi, *A New History of Modern Computing*, 258.
27 Campbell-Kelly et al., *Computer*, 417.
28 Brad A. Myers, "A Brief History of Human Computer Interaction Technology," *Interactions* 5, no. 2 (March 1998): 1, https://doi.org/10.1145/274430.274436.
29 Campbell-Kelly et al., *Computer*, 409.
30 Haigh and Ceruzzi, *A New History of Modern Computing*, 170.
31 Corey Sandler, "IBM: Colossus of Armonk," *Creative Computing* 10, no. 11 (November 1984): 298, https://www.atarimagazines.com/creative/v10n11/298 _IBM_colossus_of_Armonk.php; Campbell-Kelly et al., *Computer*, 435.
32 Haigh and Ceruzzi, *A New History of Modern Computing*, 263.
33 Isaacson, *Steve Jobs*, 404.

10. Ties That Bind

1 As Martin Campbell-Kelly et al. puts it, "The Internet sprang from a confluence of desires ... But the most romantic ideal – perhaps dating as far back as the Library of Alexandria in the ancient world – was to make readily available the world's store of knowledge." Martin Campbell-Kelly et al., *Computer: A History of*

the Information Machine, 3rd ed. (Routledge, 2014), 537, Apple Books, https:// www.taylorfrancis.com/books/mono/10.4324/9780429495373/computer -martin-campbell-kelly-william-aspray-jeffrey-yost-nathan-ensmenger.

2 Summarized in Thomas Haigh and Paul E. Ceruzzi, *A New History of Modern Computing* (MIT Press, 2021), 151, https://direct.mit.edu/books /monograph/5194/A-New-History-of-Modern-Computing.

3 Haigh and Ceruzzi, *A New History of Modern Computing*, 139, 151. See chapter 6 for a comprehensive overview.

4 Campbell-Kelly et al., *Computer*, 490–7.

5 Haigh and Ceruzzi, *A New History of Modern Computing*, chapter 6.

6 Haigh and Ceruzzi, 152.

7 Janet Abbate, *From ARPANET to Internet: A History of ARPA-Sponsored Computer Networks, 1966–1988* (University of Pennsylvania, 1994).

8 V. Rajaraman, "A Concise History of the Internet – II," *Resonance* 27, no. 12 (December 1, 2022): 2045–56, https://doi.org/10.1007/s12045-022-1505-0.

9 Scott J. Shackelford and Scott O. Bradner, *Forks in the Digital Road: Key Decisions in the History of the Internet* (Oxford University Press, 2024), 50–1.

10 Shackelford and Bradner, 52.

11 Rajaraman, "A Concise History of the Internet – II," 2054–6.

12 Shackelford and Bradner, *Forks in the Digital Road*, 87.

13 Shackelford and Bradner, 87.

14 Campbell-Kelly et al., *Computer*, 505.

15 Campbell-Kelly et al., 505.

16 Shackelford and Bradner, *Forks in the Digital Road*, 87, 89.

17 In a cavalier fashion quite disproportionate to the World Wide Web's impact, the first public announcement of the web was made via a post in the internet's hypertext newsgroup. Responding to another user, Berners-Lee said that the "World Wide Web Project aims to allow links to be made to any information anywhere," followed by "if you're interested in using the code, mail me." Walter Isaacson, *The Innovators: How a Group of Hackers, Geniuses, and Geeks Created the Digital Revolution* (Simon and Schuster, 2014), 414; Shackelford and Bradner, *Forks in the Digital Road*, 89.

18 Campbell-Kelly et al., *Computer*, 507–12 (chapter 12, section II, "The Web and Its Consequences," and section III, "Browser Wars").

19 "The Growth of the Internet, 1985–95," in *A Short History of the Internet*, National Science and Media Museum, December 3, 2020, https://www .scienceandmediamuseum.org.uk/objects-and-stories/short-history-internet.

20 "Total Number of Websites," Internet Live Stats, https://www.internetlivestats .com/total-number-of-websites/.

21 For further reading on the dotcom bubble, see Chris Gaither and Dawn C. Chmielewski, "Fears of Dot-Com Crash, Version 2.0," *Los Angeles Times*, July 16, 2006, https://www.latimes.com/archives/la-xpm-2006-jul-16-fi-overheat16-story. html; and "Early Ecommerce and the 'Dotcom Bubble,'" in *A Short History of the Internet*.

22 Gaither and Chmielewski, "Fears of Dot-Com Crash, Version 2.0."

23 Campbell-Kelly et al., *Computer*, 517–18.

24 Campbell-Kelly et al., 517.

25 Shoshana Zuboff, "Big Other: Surveillance Capitalism and the Prospects of an Information Civilization," *Journal of Information Technology* 30, no. 1 (April 4, 2015): 79, https://doi.org/10.1057/jit.2015.5.

26 Tom Seymour, Dean Frantsvog, and Satheesh Kumar, "History of Search Engines," *International Journal of Management & Information Systems* 15, no. 4 (September 2011): 47–58, https://doi.org/10.19030/ijmis.v15i4.5799.

27 Google, "Google Launches Self-Service Advertising Program," News Announcements, October 23, 2000, https://googlepress.blogspot.com/2000/10/google-launches-self-service.html.

28 Google, "Google Launches Self-Service Advertising Program."

29 Thomas Bagshaw, "The Evolution of Google AdWords – A $38 Billion Advertising Platform," *WordStream: Paid Search Marketing* (blog), December 18, 2023, https://www.wordstream.com/blog/ws/2012/06/05/evolution-of-adwords.

30 John Battelle, *The Search: How Google and Its Rivals Rewrote the Rules of Business and Transformed Our Culture*, rev. ed. (Nicholas Brealey Publishing, 2011).

31 Google, "Google Introduces New Pricing for Popular Self-Service Online Advertising Program," News Announcements, February 20, 2002, https://googlepress.blogspot.com/2002/02/google-introduces-new-pricing-for.html.

32 Google, "Google Introduces New Pricing."

33 Matthew Crain, *Profit over Privacy: How Surveillance Advertising Conquered the Internet* (University of Minnesota Press, 2021), 137.

34 Crain, *Profit over Privacy*, 141–3.

35 Steven Levy, *In the Plex: How Google Thinks, Works, and Shapes Our Lives* (Simon and Schuster, 2011), 87–93.

36 Before 2009, Google "show[ed] ads based mainly on what your interests are at a specific moment. So if you search for [digital camera] on Google, you'll get ads related to digital cameras." However, in 2009 Google announced "interest-based advertising that would ... allow advertisers to target consumers on the basis of user profiles." Crain, *Profit over Privacy*, 141–2.

37 Crain, *Profit over Privacy*, 141.

38 Zuboff, "Big Other," 79.

39 Crain, *Profit over Privacy*.

40 Zuboff, "Big Other," 79–80.

41 The Associated Press, "Google Launches Chrome Web Browser," *CBC News*, September 2, 2008, https://www.cbc.ca/news/science/google-launches-chrome-web-browser-1.710550; YouTube acquisition: Crain, *Profit over Privacy*, 139; Google Maps launch: Samuel Gibbs, "Google Maps: A Decade of Transforming the Mapping Landscape," *The Guardian*, February 8, 2015, https://www.theguardian.com/technology/2015/feb/08/google-maps-10-anniversary-iphone-android-street-view.

42 Notably, although these Google services were available in earlier years, it was not until 2012 that Google began "tracking users universally across its many services ... [to] refine its ad-targeting capabilities. Before this, Google's data collection was more of a patchwork operation wherein information was ... not connected to composite profiles." Crain, *Profit over Privacy*, 142.

43 Zuboff, "Big Other," 83.

44 Zuboff, 79.

45 Recounting a 2002 conversation with Google's cofounder Larry Page, Kevin Kelly explains his confusion surrounding Google's motivations for providing free web search. Kelly said, "Larry, I still don't get it. There are so many search companies. Web search, for free? Where does that get you?" Page's response: "Oh, we're really making an AI." Kevin Kelly, "The Three Breakthroughs That Have Finally Unleashed AI on the World," *Wired*, October 27, 2014, https://

www.wired.com/2014/10/future-of-artificial-intelligence/; Adam Fisher, *Valley of Genius: The Uncensored History of Silicon Valley (As Told by the Hackers, Founders, and Freaks Who Made It Boom)* (Twelve, 2018), 278.

46 Rachel Hespell, "Our 10 Biggest AI Moments So Far," *The Keyword* (Google blog), September 26, 2023, https://blog.google/technology/ai/google-ai-ml-timeline/; Steven Levy, "How Google Is Remaking Itself as a 'Machine Learning First' Company," *Wired*, June 22, 2016, https://www.wired.com/2016/06/how-google-is-remaking-itself-as-a-machine-learning-first-company/.

47 Hespell, "Our 10 Biggest AI Moments"; Levy, "How Google Is Remaking Itself."

48 Hespell; Levy.

49 Crain, *Profit over Privacy*, 139.

50 Crain, 140.

11. The New Machine

1 Walter Isaacson, *Steve Jobs* (Simon and Schuster, 2011), 591.

2 Isaacson, *Steve Jobs*, 293, 317–18; Martin Campbell-Kelly et al., *Computer: A History of the Information Machine*, 3rd ed. (Routledge, 2014), 524, Apple Books, https://www.taylorfrancis.com/books/mono/10.4324/9780429495373/computer-martin-campbell-kelly-william-aspray-jeffrey-yost-nathan-ensmenge.

3 Isaacson, *Steve Jobs*, chapter 24, "Restoration."

4 Isaacson, *Steve Jobs*, 400.

5 Isaacson, 403.

6 Isaacson, 443.

7 Bill Murphy Jr., "27 Years Ago, Steve Jobs Explained How He Fired People. Here's How He Did It," *Inc.*, December 15, 2022, https://www.inc.com/bill-murphy-jr/27-years-ago-steve-jobs-explained-how-he-fired-people-heres-how-he-did-it.html.

8 Isaacson, *Steve Jobs*, 445.

9 Campbell-Kelly et al., *Computer*, 524.

10 Walter Isaacson, "The Real Leadership Lessons of Steve Jobs," *Harvard Business Review*, April 2012, https://hbr.org/2012/04/the-real-leadership-lessons-of-steve-jobs.

11 Isaacson, *Steve Jobs*, 608.

12 Isaacson, *Steve Jobs*, 608.

13 Referencing the possibility of mobile phone makers adding music players to their devices, Steve Jobs himself said, "If we don't cannibalize [the iPod] ourselves, someone else will." Quoted in Isaacson, "The Real Leadership Lessons of Steve Jobs."

14 The "iTunes Phone" was so notoriously bad that *Wired* summarized the disappointment on its November 2005 cover: "YOU CALL *THIS* THE PHONE OF THE FUTURE?"; Isaacson, *Steve Jobs*, 609; Fred Vogelstein, "The Untold Story: How the iPhone Blew Up the Wireless Industry," *Wired*, January 9, 2008, https://www.wired.com/2008/01/ff-iphone/.

15 Brian Merchant, "The Secret Origin Story of the iPhone," *The Verge*, June 13, 2017, https://www.theverge.com/2017/6/13/15782200/one-device-secret-history-iphone-brian-merchant-book-excerpt.

16 Apple, "Apple, Motorola & Cingular Launch World's First Mobile Phone with iTunes," Apple Newsroom, press release, September 7, 2005, https://www.apple.com/newsroom/2005/09/07Apple-Motorola-Cingular-Launch-Worlds-First-Mobile-Phone-with-iTunes/.

17 *Steve Jobs Presents the iTunes Phone*, YouTube, 2009, https://www.youtube.com /watch?v=TWSRgsk2oaw.
18 Isaacson, *Steve Jobs*, 612.
19 Merchant, "The Secret Origin Story of the iPhone."
20 Merchant.
21 Vogelstein, "The Untold Story"; Jonathan Weber, "Apple to Join Acorn, VLSI in Chip-Making Venture," *Los Angeles Times*, November 28, 1990, https://www .latimes.com/archives/la-xpm-1990-11-28-fi-4993-story.html.
22 Jim Ledin and Dave Farley, *Modern Computer Architecture and Organization: Learn X86, ARM, and RISC-V Architectures and the Design of Smartphones, PCs, and Cloud Servers* (Packt Publishing, 2022), 10–14.
23 Merchant, "The Secret Origin Story of the iPhone."
24 Jake Swearingen, "How Steve Jobs Faked His Way through Unveiling the iPhone," *Intelligencer*, January 9, 2017, https://nymag.com/intelligencer/2017/01/how -steve-jobs-faked-his-way-through-unveiling-the-iphone.html.
25 Swearingen, "How Steve Jobs Faked His Way."
26 Jeannine Mancini, "Steve Jobs Rigged the First iPhone Demo by Faking Full Signal Strength and Secretly Swapping Devices Because of Fragile Prototypes and Bug-Riddled Software – The Engineers Were So Nervous They Got Drunk during Presentation to Calm Their Nerves," *Yahoo Finance*, December 15, 2023, https:// finance.yahoo.com/news/steve-jobs-rigged-first-iphone-152527272.html.

12. The End Game

1 ImageNet, www.image-net.org.
2 Wayne E. Carlson, "Industrial Light and Magic (ILM)," in *Computer Graphics and Computer Animation: A Retrospective Overview* (Ohio State University, 2017), https:// ohiostate.pressbooks.pub/graphicshistory/chapter/11-2-industrial-light-and -magic-ilm/; Martin Campbell-Kelly et al., *Computer: A History of the Information Machine* (Routledge, 2014), 582, Apple Books, https://www.taylorfrancis.com /books/mono/10.4324/9780429495373/computer-martin-campbell-kelly -william-aspray-jeffrey-yost-nathan-ensmenge; Walter Isaacson, *Steve Jobs* (Simon and Schuster, 2011), 317–19.
3 Carlson, "Industrial Light and Magic (ILM)"; Campbell-Kelly, *Computer*, 582; Isaacson, *Steve Jobs*, 317–19.
4 "A Short History of CGI," *3Dlines* (blog), May 22, 2024, https://web.archive.org /web/20240913053953/https://www.3dlines.co.uk/a-short-history-of-cgi/.
5 Jim Ledin and Dave Farley, *Modern Computer Architecture and Organization: Learn X86, ARM, and RISC-V Architectures and the Design of Smartphones, PCs, and Cloud Servers* (Packt Publishing, 2022), chapter 3.
6 "CPU vs. GPU: What's the Difference?," Intel, https://www.intel.com/content /www/us/en/products/docs/processors/cpu-vs-gpu.html.
7 Ledin and Farley, *Modern Computer Architecture and Organization*.
8 Steven Levy, "Fei-Fei Li Started an AI Revolution by Seeing Like an Algorithm," *Wired*, accessed January 9, 2025, https://www.wired.com/story/plaintext-fei-fei-li -ai-revolution-seeing-imagenet-algorithm/.
9 Michael J. Wooldridge, *A Brief History of Artificial Intelligence: What It Is, Where We Are, and Where We Are Going* (Flatiron Books, 2021), 206–7, Apple Books.
10 Jia Deng et al., "ImageNet: A Large-Scale Hierarchical Image Database," in *2009 IEEE Conference on Computer Vision and Pattern Recognition* (IEEE, 2009), 248–55.

11 Rony Chow, "ImageNet: A Pioneering Vision for Computers," History of Data Science, August 27, 2021, https://www.historyofdatascience.com/imagenet-a-pioneering-vision-for-computers/; "ImageNet Large Scale Visual Recognition Challenge (ILSVRC)," ImageNet, https://www.image-net.org/challenges/LSVRC/index.php.

12 Alex Krizhevsky, Ilya Sutskever, and Geoffrey E. Hinton, "ImageNet Classification with Deep Convolutional Neural Networks," in *Advances in Neural Information Processing Systems*, vol. 25 (Curran Associates, 2012), https://proceedings.neurips.cc/paper_files/paper/2012/hash/c399862d3b9d6b76c8436e924a68c45b-Abstract.html.

13 Wooldridge, *A Brief History of Artificial Intelligence*, 202.

14 Michael A. Nielsen, *Neural Networks and Deep Learning* (Determination Press, 2015), http://neuralnetworksanddeeplearning.com.

15 Wooldridge, *A Brief History of Artificial Intelligence*, 206.

16 Siddhesh Bangar, "AlexNet Architecture Explained," *Medium* (blog), June 24, 2022, https://medium.com/@siddheshb008/alexnet-architecture-explained-b6240c528bd5.

17 James Briggs and Laura Carnevali, *Embedding Methods for Image Search* (Pinecone, n.d.), chapter 3, accessed January 9, 2025, https://www.pinecone.io/learn/series/image-search/.

18 Briggs and Carnevali, *Embedding Methods*, chapter 3.

19 Association for Computing Machinery (ACM), "Fathers of the Deep Learning Revolution Receive 2018 ACM A.M. Turing Award," ACM, March 27, 2019, https://www.acm.org/media-center/2019/march/turing-award-2018.

20 Cade Metz, "'The Godfather of A.I.' Leaves Google and Warns of Danger Ahead," *New York Times*, May 1, 2023, sec. Technology, https://www.nytimes.com/2023/05/01/technology/ai-google-chatbot-engineer-quits-hinton.html.

14. Aftermath

1 Michael Wooldridge, *A Brief History of Artificial Intelligence: What It Is, Where We Are, and Where We Are Going* (Flatiron Books, 2021), chapter 5 and 188, Apple Books.

2 Wooldridge, *A Brief History of Artificial Intelligence*, 189.

3 Wooldridge, 189.

4 In his article "The Three Breakthroughs That Have Finally Unleashed AI on the World," the founding editor of *Wired*, Kevin Kelly, recounts a 2002 conversation with Google's Larry Page. Asked why Google would provide its search engine for free, Page responded, "Oh, we're really making an AI." Kevin Kelly, "The Three Breakthroughs That Have Finally Unleashed AI on the World," *Wired*, October 27, 2014, https://www.wired.com/2014/10/future-of-artificial-intelligence/.

5 Transcript and audio from "Alphabet Q1 2023 Earnings Call," Alphabet Investor Relations, April 25, 2023, https://abc.xyz/investor/events/2023-q1-earnings-call.

6 Steven Levy, "Secret of Googlenomics: Data-Fueled Recipe Brews Profitability," *Wired*, May 22, 2009, https://www.wired.com/2009/05/nep-googlenomics/.

7 Rachel Hespell, "A Timeline of Google's Biggest AI and ML Moments," *The Keyword* (Google blog), September 26, 2023, https://blog.google/technology/ai/google-ai-ml-timeline/.

8 Rony Chow, "Google Brain: The Brains behind Your Search Engine," History of Data Science, June 3, 2021, https://www.historyofdatascience.com/google

-brain-the-brains-behind-your-search-engine/; Anna Crowley Redding, *Google It: A History of Google* (Feiwel & Friends, 2018), 283, Apple Books.

9 "Our Mission: Be the World's Most Trusted Driver," Waymo, accessed October 4, 2024, https://web.archive.org/web/20241004135614/https://waymo.com /about/.

10 "Google X: Timeline," X, The Moonshot Factory, accessed December 12, 2024, https://web.archive.org/web/20241227062922/https://x.company/; John Markoff, "Google Puts Money on Robots, Using the Man behind Android," *New York Times*, December 4, 2013, https://www.nytimes.com/2013/12/04/ technology/google-puts-money-on-robots-using-the-man-behind-android.html; Redding, *Google It*, Part Three, 16–17.

11 Redding, *Google It*, 247.

12 Steven Levy, *In the Plex: How Google Thinks, Works, and Shapes Our Lives* (Simon and Schuster, 2011), 235–6; Chris Velazco, "How Google's Smartphones Have Evolved Since 2007," *Engadget*, October 3, 2017, https://www.engadget.com/2017-10-03-a -look-back-at-googles-smartphones.html.

13 Larry Page, "We've Acquired Motorola Mobility," *The Keyword* (Google blog), May 22, 2012, https://blog.google/inside-google/company-announcements/weve -acquired-motorola-mobility/.

14 Christina Bonnington, "Google Buys Nest for $3.2 Billion in Cash," *Wired*, January 13, 2014, https://www.wired.com/2014/01/google-nest-buy/.

15 Redding, *Google It*, 328.

16 Redding, 332.

17 "U of T Neural Networks Start-up Acquired by Google," University of Toronto, media release, March 12, 2013, https://media.utoronto.ca/media-releases/u -of-t-neural-networks-start-up-acquired-by-google/; "Google Buys University of Toronto Startup," *CBC News*, March 13, 2013, https://www.cbc.ca/news/science /google-buys-university-of-toronto-startup-1.1373641.

18 "U of T Neural Networks Start-up"; "Google Buys University of Toronto Startup."

19 Wooldridge, *A Brief History of Artificial Intelligence*, chapter 5 and 187.

20 Wooldridge, chapter 5.

21 Alyson Shontell, "Snapchat Buys Looksery, a 2-Year-Old Startup That Lets You Photoshop Your Face While You Video Chat," *Business Insider*, September 15, 2015, https://www.businessinsider.com/snapchat-buys-looksery-2015-9; Anweshbiswas, "Snapchat's History of Successful AI & AR Usage," *Medium* (blog), October 21, 2023, https://medium.com/@work.anweshbiswas/snapchats-history-of-successful -ai-ar-usage-54076103d1fc.

22 "AI on Snapchat: Improved Transparency, Safety, and Policies," Snapchat Privacy, Safety, and Policy Hub, April 16, 2024, https://values.snap.com/news/ai-on -snapchat-improved-transparency-safety-policy.

23 "About: Epagogix," DBpedia, https://dbpedia.org/page/Epagogix.

24 Malcolm Gladwell, "The Formula," *The New Yorker*, October 8, 2006, https://www .newyorker.com/magazine/2006/10/16/the-formula; Jer Thorp, *Living in Data: A Citizen's Guide to a Better Information Future* (Macmillan, 2021); Jeffrey Wells, "Epagogix," *Hollywood Elsewhere* (blog), October 26, 2006, https://hollywood -elsewhere.com/epagogix/.

25 A.J. Chisling, "AI Can Approve Film Scripts on Its Own – and Other Fun Algorithm Facts," *Medium* (blog), January 10, 2018, https://avajoy.medium.com/ai-can -approve-film-scripts-on-its-own-and-other-fun-algorithm-facts-51222b97f977.

26 Kanadpriya Basu et al., "Artificial Intelligence: How Is It Changing Medical Sciences and Its Future?," *Indian Journal of Dermatology* 65, no. 5 (2020): 365–70, https://doi.org/10.4103/ijd.IJD_421_20; Salman Bahoo et al., "Artificial Intelligence in Finance: A Comprehensive Review through Bibliometric and Content Analysis," *SN Business & Economics* 4, no. 2 (January 20, 2024): 23, https://doi.org/10.1007/s43546-023-00618-x.

27 "First Computer Film Critic," Guinness World Records, 2003, https://www.guinnessworldrecords.com/world-records/100881-first-computer-film-critic.html.

15. GPT

1 Jaclyn Peiser, "The Rise of the Robot Reporter," *New York Times*, February 5, 2019, sec. Business, https://www.nytimes.com/2019/02/05/business/media/artificial-intelligence-journalism-robots.html.

2 Tom B. Brown et al., "Language Models Are Few-Shot Learners" (arXiv, July 22, 2020), https://doi.org/10.48550/arXiv.2005.14165.

3 Bernard Marr, "A Short History of ChatGPT: How We Got to Where We Are Today," *Forbes*, May 19, 2023, https://www.forbes.com/sites/bernardmarr/2023/05/19/a-short-history-of-chatgpt-how-we-got-to-where-we-are-today/; OpenAI, "GPT-2: 1.5B Release," OpenAI news release, November 5, 2019, https://openai.com/index/gpt-2-1-5b-release/.

4 OpenAI, "Introducing ChatGPT," November 30, 2022, https://openai.com/index/chatgpt/.

5 Chuan Li, "OpenAI's GPT-3 Language Model: A Technical Overview," *Lambda* (blog), June 3, 2020, https://lambdalabs.com/blog/demystifying-gpt-3.

6 Cade Metz, *Genius Makers: The Mavericks Who Brought AI to Google, Facebook, and the World* (Penguin Random House, 2021), prologue, 2, 98; "Google Buys University of Toronto Startup," *CBC News*, March 13, 2013, https://www.cbc.ca/news/science/google-buys-university-of-toronto-startup-1.1373641; Melanie Mitchell, *Artificial Intelligence: A Guide for Thinking Humans* (Farrar, Straus and Giroux, 2019), 87.

7 Metz, *Genius Makers*, prologue, 142–4; "Ilya Sutskever's Home Page," https://www.cs.toronto.edu/~ilya/.

8 Rachel Metz, "Ilya Sutskever, Co-Founder and Chief Scientist, Leaves OpenAI," *TIME*, May 15, 2024, https://time.com/6978195/ilya-sutskever-leaves-open-ai/.

9 Krystal Hu and Kenrick Cai, "Exclusive: OpenAI to Remove Non-Profit Control and Give Sam Altman Equity," *Reuters*, September 26, 2024, sec. Artificial Intelligence, https://www.reuters.com/technology/artificial-intelligence/openai-remove-non-profit-control-give-sam-altman-equity-sources-say-2024-09-25/; Cade Metz, "OpenAI Details Plans for Becoming a For-Profit Company," *New York Times*, December 27, 2024, sec. Technology, https://www.nytimes.com/2024/12/27/technology/openai-public-benefit-corporation.html.

10 "Our Structure," OpenAI, https://openai.com/our-structure/.

11 "OpenAI Charter," OpenAI, https://openai.com/charter/.

12 Trapit Bansal et al., "Competitive Self-Play," *OpenAI: Milestones*, October 11, 2017, https://openai.com/index/competitive-self-play/; OpenAI, *Competitive Self-Play*, YouTube, 2017, https://www.youtube.com/watch?v=OBcjhp4KSgQ.

13 David Silver et al., "AlphaZero: Shedding New Light on Chess, Shogi, and Go," Google DeepMind, December 6, 2018, https://deepmind.google/discover /blog/alphazero-shedding-new-light-on-chess-shogi-and-go/; James Somers, "How the Artificial Intelligence Program AlphaZero Mastered Its Games," *The New Yorker*, December 28, 2018, https://www.newyorker.com/science/elements /how-the-artificial-intelligence-program-alphazero-mastered-its-games.

14 Silver et al., "AlphaZero"; Somers, "How the Artificial Intelligence Program"; Mitchell, *Artificial Intelligence*, 160.

15 David Silver et al., "A General Reinforcement Learning Algorithm That Masters Chess, Shogi, and Go through Self-Play," *Science* 362, no. 6419 (December 7, 2018): 27, https://doi.org/10.1126/science.aar6404.

16 Silver et al., "A General Reinforcement Learning Algorithm," 4, 27.

17 Silver et al., "AlphaZero."

18 Ashish Vaswani et al., "Attention Is All You Need" (arXiv, June 12, 2017), https:// doi.org/10.48550/arXiv.1706.03762.

19 Mitchell, *Artificial Intelligence*, 184–6.

20 Mitchell, 174–82.

21 Mitchell, 178.

22 Rewon Child and Scott Gray, "Generative Modeling with Sparse Transformers," OpenAI, April 23, 2019, https://openai.com/index/sparse-transformer/.

23 Metz, *Genius Makers*, 163.

24 Alec Radford et al., "Improving Language Understanding by Generative Pre-Training," June 2018, https://cdn.openai.com/research-covers/language -unsupervised/language_understanding_paper.pdf.

25 Alec Radford et al., "Language Models Are Unsupervised Multitask Learners," OpenAI, 2019, https://cdn.openai.com/better-language-models/language_models _are_unsupervised_multitask_learners.pdf; Alec Radford et al., "Better Language Models and Their Implications," OpenAI, February 14, 2019, https://openai.com /index/better-language-models/; "GPT-2: 1.5B Release," OpenAI, November 5, 2019, https://openai.com/index/gpt-2-1-5b-release/.

26 Marr, "A Short History of ChatGPT."

27 "GPT-2: 1.5B Release."

28 Li, "OpenAI's GPT-3 Language Model"; Brown et al., "Language Models Are Few-Shot Learners."

29 "OpenAI API," OpenAI, June 11, 2020, https://openai.com/index/openai -api/; "OpenAI's API Now Available with No Waitlist," OpenAI, November 18, 2021, https://openai.com/index/api-no-waitlist/; David Eliot and Rod Bantjes, "Climate Science vs Denial Machines: How AI Could Manufacture Scientific Authority for Far-Right Disinformation," in *Political Ecologies of the Far Right: Fanning the Flames*, ed. Irma Kinga Allen et al. (Manchester University Press, 2024), 213–31, https://www.researchgate.net/publication/380931631_Climate _science_vs_denial_machines_How_AI_could_manufacture_scientific _authority_for_far-right_disinformation; Will Douglas Heaven, "OpenAI's New Language Generator GPT-3 Is Shockingly Good – and Completely Mindless," *MIT Technology Review*, July 20, 2020, https://www.technologyreview .com/2020/07/20/1005454/openai-machine-learning-language-generator-gpt-3 -nlp/; Brown et al., "Language Models Are Few-Shot Learners."

30 "OpenAI API."

31 Ashley Pilipiszyn and OpenAI, "GPT-3 Powers the Next Generation of Apps," OpenAI, March 25, 2021, https://openai.com/index/gpt-3-apps/.

32 "New GPT-3 Capabilities: Edit & Insert," OpenAI, March 15, 2022, https://openai.com/index/gpt-3-edit-insert/; "Introducing ChatGPT."

33 David Eliot and David Murakami Wood, "Google and Microsoft Are Creating a Monopoly on Coding in Plain Language," *The Conversation*, September 8, 2021, http://theconversation.com/google-and-microsoft-are-creating-a-monopoly-on-coding-in-plain-language-166258; Richard Gao, "How to Use ChatGPT to Write Code and Generate Websites," *Medium* (blog), December 11, 2022, https://rich-gaogle.medium.com/using-gpt-3-to-write-code-337d80b3647f.

34 "DALL·E: Creating Images from Text," OpenAI, January 5, 2021, https://openai.com/index/dall-e/.

35 Boris Dayma et al., "DALL-E Mini Explained," Weights & Biases, July 2021, https://wandb.ai/dalle-mini/dalle-mini/reports/DALL-E-Mini-Explained–Vmlldzo4NjIxODA; Will Knight, "Inside DALL-E Mini, the Internet's Favorite AI Meme Machine," *Wired*, June 27, 2022, https://www.wired.com/story/dalle-ai-meme-machine/.

36 "Introducing ChatGPT."

37 "GPT-4 Is OpenAI's Most Advanced System, Producing Safer and More Useful Responses," OpenAI, March 14, 2023, https://openai.com/index/gpt-4/; "ChatGPT – Release Notes," OpenAI Help Center, https://help.openai.com/en/articles/6825453-chatgpt-release-notes.

16. Beyond Generative AI

1 Andrew McDiarmid, "Hyping Artificial Intelligence Hinders Innovation," episode 163, *How AI Changed – in a Very Big Way – Around 2000* (podcast), Mind Matters, December 7, 2021, https://mindmatters.ai/2021/12/how-ai-changed-in-a-very-big-way-around-the-year-2000/.

2 "AI in Mammography: Improving Breast Cancer Screening with Artificial Intelligence," Google Health, https://health.google/caregivers/mammography/; Andreas S. Panayides et al., "AI in Medical Imaging Informatics: Current Challenges and Future Directions," *IEEE Journal of Biomedical and Health Informatics* 24, no. 7 (2020): 1837–57, https://doi.org/10.1109/JBHI.2020.2991043; Abid Haleem, Mohd Javaid, and Ibrahim Haleem Khan, "Current Status and Applications of Artificial Intelligence (AI) in Medical Field: An Overview," *Current Medicine Research and Practice* 9, no. 6 (2019): 231–7, https://doi.org/10.1016/j.cmrp.2019.11.005.

3 Jordan Burrows, "IRS Using Artificial Intelligence to Make Sure People Aren't Playing the System," *CBS News*, March 27, 2024, https://www.cbsnews.com/detroit/news/irs-using-artificial-intelligence-ai-taxes/; U.S. Government Accountability Office, "Artificial Intelligence May Help IRS Close the Tax Gap," *WatchBlog* (blog), June 6, 2024, https://www.gao.gov/blog/artificial-intelligence-may-help-irs-close-tax-gap.

4 Rosella (Qian-Ze) Zhu, "A Sky Full of Data: Weather Forecasting in the Age of AI," *Science in the News*, March 4, 2024, https://sites.harvard.edu/sitn/2024/03/04/ai_weather_forecasting/; Dmitrii Kochkov et al., "Neural General Circulation Models for Weather and Climate," *Nature* 632, no. 8027 (August 29, 2024): 1060–6, https://doi.org/10.1038/s41586-024-07744-y; Ilan Price et al., "GenCast:

Diffusion-Based Ensemble Forecasting for Medium-Range Weather" (arXiv, December 25, 2023), https://doi.org/10.48550/arXiv.2312.15796.

5 Valentina Bellini et al., "Artificial Intelligence in Operating Room Management," *Journal of Medical Systems* 48 (2024): article 19, https://doi.org/10.1007/s10916 -024-02038-2.

6 Sam Becker, "US Farms Are Making an Urgent Push into AI. It Could Help Feed the World," *BBC News*, March 27, 2024, https://www.bbc.com/worklife /article/20240325-artificial-intelligence-ai-us-agriculture-farming.

7 Julian Murphy, "Chilling: The Constitutional Implications of Body-Worn Cameras and Facial Recognition Technology at Public Protests," *Washington and Lee Law Review Online* 75, no. 1 (August 30, 2018): article 1, https://scholarlycommons .law.wlu.edu/wlulr-online/vol75/iss1/1; Pete Fussey, Bethan Davies, and Martin Innes, "'Assisted' Facial Recognition and the Reinvention of Suspicion and Discretion in Digital Policing," *The British Journal of Criminology* 61, no. 2 (March 1, 2021): 325–44, https://doi.org/10.1093/bjc/azaa068.

8 Paul Mozur and Adam Satariano, "A.I. Begins Ushering In an Age of Killer Robots," *New York Times*, July 2, 2024, sec. Technology, https://www.nytimes .com/2024/07/02/technology/ukraine-war-ai-weapons.html.

9 Hope Reese, "What Happens When Police Use AI to Predict and Prevent Crime?," *JSTOR Daily*, February 23, 2022, https://daily.jstor.org/what-happens-when -police-use-ai-to-predict-and-prevent-crime/; EUCPN and Majsa Storbeck, *Artificial Intelligence and Predictive Policing: Risks and Challenges* (EUCPN, 2022), https://eucpn.org/sites/default/files/document/files/PP%20%282%29.pdf.

10 Reinhard Klette, *Concise Computer Vision: An Introduction into Theory and Algorithms* (Springer Science & Business Media, 2014).

11 "What Is Computer Vision?" IBM: Think, July 27, 2021, https://www.ibm.com /think/topics/computer-vision.

17. The New Computer

1 John Tinnell, "The Philosopher of Palo Alto," *OUPblog* (blog), December 9, 2017, https://blog.oup.com/2017/12/palo-alto-philosophy-weiser-technology/.

2 John Tinnell, *The Philosopher of Palo Alto: Mark Weiser, Xerox PARC, and the Original Internet of Things* (University of Chicago Press, 2023), 61.

3 John Markoff, "Mark Weiser, a Leading Computer Visionary, Dies at 46," *New York Times*, May 1, 1999, sec. Business, https://www.nytimes.com/1999/05/01 /business/mark-weiser-a-leading-computer-visionary-dies-at-46.html.

4 Tinnell, *The Philosopher of Palo Alto*, 41.

5 Martin Heidegger, *Being and Time: A Translation of Sein Und Zeit* (State University of New York Press, 1927); Tinnell, *The Philosopher of Palo Alto*, 54–60.

6 Marc Weiser, "The World Is Not a Desktop," *Interactions* 1, no. 1 (January 2, 1994): 7–8, https://doi.org/10.1145/174800.174801.

7 Tinnell, *The Philosopher of Palo Alto*, 77–8.

8 Markoff, "Mark Weiser."

9 Tinnell, *The Philosopher of Palo Alto*, 101.

10 In Weiser's obituary, the example of the cane is credited to Heidegger; however, there is no evidence that Heidegger ever wrote about the blind man's cane, and the lessons drawn from the example seem to suggest that Weiser was inspired

by Merleau-Ponty. Yet, Heidegger's theories regarding human-technology interaction (ready-to-hand) are often discussed in reference to Merleau-Ponty's example of the blind man's cane, and it is possible that they had influence on Weiser as well. See Markoff, "Mark Weiser."

11 Markoff, 52.

12 Mark Weiser and John Seely Brown, "Designing Calm Technology" (Xerox PARC, December 21, 1995), https://people.csail.mit.edu/rudolph/Teaching/weiser .pdf; Mark Weiser, "The Computer for the 21st Century," *ACM SIGMOBILE Mobile Computing and Communications Review* 3, no. 3 (July 1, 1999): 3–11, https://doi .org/10.1145/329124.329126; Tinnell, *The Philosopher of Palo Alto*, 102–3.

13 Markoff, "Mark Weiser."

14 Markoff.

15 Tinnell, "The Philosopher of Palo Alto."

16 Amy Goetzman, "Mark Weiser and the Origins of the Internet of Things," Connector Supplier, September 12, 2023, https://connectorsupplier.com/mark -weiser-and-the-origins-of-the-internet-of-things/.

17 Shannon Mattern, *Code and Clay, Data and Dirt: Five Thousand Years of Urban Media* (University of Minnesota Press, 2017), https://www.upress.umn.edu /9781517902445/code-and-clay-data-and-dirt/; David Murakami Wood and Debra Mackinnon, "Partial Platforms and Oligoptic Surveillance in the Smart City," *Surveillance & Society* 17, no. 1/2 (March 31, 2019): 176–82, https://doi .org/10.24908/ss.v17i1/2.13116; David Murakami Wood, "Urban Surveillance After the End of Globalization," in *A Research Agenda for Cities*, ed. John Rennie Short (Edward Elgar Publishing, 2017), 38–50, https://www.researchgate.net /publication/370810772_Urban_surveillance_after_the_end_of_globalization.

18 James Darley, "Top 10: Smart Cities," *Sustainability Magazine*, December 4, 2024, https://sustainabilitymag.com/top10/top-10-smart-cities; Maya Derrick, "Top 10: Smart Cities," *Energy Digital*, July 31, 2024, https://energydigital.com/top10/top -10-smart-city.

19 Monique Mann et al., "#BlockSidewalk to Barcelona: Technological Sovereignty and the Social License to Operate Smart Cities," *Journal of the Association for Information Science and Technology* 71, no. 9 (2020): 1103–15, https://doi .org/10.1002/asi.24387.

20 Weiser, "The Computer for the 21st Century."

21 Weiser, 5.

22 Weiser, 3.

23 Tinnell, "The Philosopher of Palo Alto."

24 Tinnell, *The Philosopher of Palo Alto*, 292–3.

25 Tinnell, *The Philosopher of Palo Alto*, 106.

26 For further reading, see Tinnell, chapter 4.

27 Cathy Hackl, "On-Device AI and the Voice-Powered Future of Computing," *Forbes*, November 14, 2024, sec. CMO Network, https://www.forbes.com/sites /cathyhackl/2024/11/14/on-device-ai-the-voice-powered-future-of-computing/.

28 Rob Wile, "Why Everyone Is Suddenly Talking about Nvidia, the Nearly $3 Trillion-Dollar Company Fueling the AI Revolution," *NBC News*, February 24, 2024, sec. Business News, https://www.nbcnews.com/business/business-news /what-is-nvidia-what-do-they-make-ai-artificial-intelligence-rcna140171.

29 Daniel Bourke, "Apple's New M1 Chip Is a Machine Learning Beast," *Medium: TDS Archive* (blog), December 24, 2020, https://medium.com/towards-data -science/apples-new-m1-chip-is-a-machine-learning-beast-70ca8bfa6203.

30 Nick Bilton, "Why Google Glass Broke," *New York Times*, February 4, 2015, sec. Style, https://www.nytimes.com/2015/02/05/style/why-google-glass-broke.html; Steven John, "Google Glass: A History of the Discontinued Smart Glasses, What They Did, Why They Failed," *Business Insider*, May 13, 2024, https://www.businessinsider.com/google-glass.

31 Anna Crowley Redding, *Google It: A History of Google* (Feiwel & Friends, 2018), chapter 18.

32 Erik Sherman, "Google Glass: $1,500 to Buy, $80 to Make?," *CBS News*, May 1, 2014, https://web.archive.org/web/20180104013857/https://www.cbsnews.com/news/google-glass-1500-to-buy-80-to-make/.

33 Josh Lowensohn, "Unlocked iPhone 5 Could Arrive in Apple's Web Store Tonight," *CNET News*, November 29, 2012, https://web.archive.org/web/20130329053053/http://news.cnet.com/8301-13579_3-57556310-37/unlocked-iphone-5-could-arrive-in-apples-web-store-tonight/; Hayley Tsukayama, "Apple Begins Selling Unlocked iPhone 5," *Washington Post*, November 30, 2012, https://www.washingtonpost.com/business/technology/apple-begins-selling-unlocked-iphone-5/2012/11/30/0d47bd7a-3aeb-11e2-8a97-363b0f9a0ab3_story.html.

34 Jason Hong, "Considering Privacy Issues in the Context of Google Glass," *Communications of the ACM* 56, no. 11 (November 1, 2013): 10–11, https://doi.org/10.1145/2524713.2524717.

35 Redding, *Google It*, 201; Brian Lystgaard Due, "The Social Construction of a Glasshole: Google Glass and Multiactivity in Social Interaction," *PsychNology Journal* 13, no. 2–3 (2016): 149–78.

36 Tim Bradshaw, "Google Bets on 'Internet of Things' with $3.2bn Nest Deal," *Financial Times*, January 13, 2014, https://www.ft.com/content/90b8714a-7c99-11e3-b514-00144feabdc0; Alexei Oreskovic and Poornima Gupta, "Google Gains Entry to Home and Prized Team with $3.2 Billion Nest Deal," *Reuters*, January 14, 2014, sec. Technology, https://www.reuters.com/article/technology/google-gains-entry-to-home-and-prized-team-with-32-billion-nest-deal-idUSBREA0C1HP/.

37 Andrew Gebhart, "Google and Nest Combine into a New Smart Home Brand," *CNET*, May 7, 2019, https://www.cnet.com/home/smart-home/google-and-nest-combine-into-a-new-smart-home-brand/.

38 Steven Levy, "Google Glass 2.0 Is a Startling Second Act," *Wired*, July 18, 2017, https://www.wired.com/story/google-glass-2-is-here/; Redding, *Google It*, 202.

39 Mitchell Clark, "Google Glass Enterprise Edition Is No More," *The Verge*, March 15, 2023, https://www.theverge.com/2023/3/15/23641872/google-glass-enterprise-edition-discontinued-support.

40 Sarah McBride, "With HoloLens, Microsoft Aims to Avoid Google's Mistakes," *Reuters*, May 23, 2016, sec. Technology, https://www.reuters.com/article/technology/with-hololens-microsoft-aims-to-avoid-googles-mistakes-idUSKCN0YE1LZ/; Gurmeet Singh Pandher, "Microsoft HoloLens Preorders: Price, Specs of the Augmented Reality Headset," The Bitbag, March 3, 2016, https://web.archive.org/web/20160304102828/http://www.thebitbag.com/microsoft-hololens-preorders-price-specs-of-the-augmented-reality-headset/137410.

41 Sarah McBride, "With HoloLens, Microsoft Aims to Avoid Google's Mistakes."

42 Heather Kelly, "Microsoft's New $3,500 HoloLens 2 Headset Means Business," *CNN*, February 24, 2019, https://www.cnn.com/2019/02/24/tech/microsoft-hololens-2/index.html.

43 Deborah Bach, "U.S. Army to Use HoloLens Technology in High-Tech Headsets for Soldiers," *Microsoft News*, June 8, 2021, https://news.microsoft.com/source

/features/digital-transformation/u-s-army-to-use-hololens-technology-in-high
-tech-headsets-for-soldiers/; Umar Shakir, "US Army Orders More Microsoft AR
Headsets Now That They No Longer Make Soldiers Want to Barf," *The Verge*,
September 13, 2023, https://www.theverge.com/2023/9/13/23871859/us-army
-microsoft-ivas-ar-goggles-success-new-contract-hololens.

44 David Pierce, "Pixel by Pixel: How Google Is Trying to Focus and Ship the
Future," *The Verge*, May 11, 2022, https://www.theverge.com/23065820/google
-io-ambient-computing-pixel-android-phones-watches-software; Hector Ouilhet,
"Human-Centered Ambient Home Tech," Google Design, November 12, 2020,
https://design.google/library/mastering-ambiance-ambient-computing.

45 "Alphabet Q3 2019 Earnings Call," Alphabet Investor Relations, October 28, 2019,
https://abc.xyz/assets/investor/static/pdf/2019_Q3_Earnings_Transcript.pdf.

46 *Made by Google '19*, YouTube, October 15, 2019, 03:04–03:44, https://www.youtube
.com/watch?v=XKmsYB54zBk.

47 Nicholas Confessore, "Cambridge Analytica and Facebook: The Scandal and
the Fallout So Far," *New York Times*, April 4, 2018, sec. U.S., https://www.nytimes
.com/2018/04/04/us/politics/cambridge-analytica-scandal-fallout.html; Georgia
Wells, Jeff Horwitz, and Deepa Seetharaman, "Facebook Knows Instagram Is Toxic
for Teen Girls, Company Documents Show," *Wall Street Journal*, September 14,
2021, sec. Technology, https://www.wsj.com/articles/facebook-knows-instagram
-is-toxic-for-teen-girls-company-documents-show-11631620739.

48 Sundar Pichai, "Making AI Work for Everyone," *The Keyword* (Google blog), May 17,
2017, https://blog.google/technology/ai/making-ai-work-for-everyone/; Anita
Balakrishnan, "Google CEO Cites Shift from a Mobile World to an AI World," *CNBC*,
April 28, 2016, sec. Technology, https://www.cnbc.com/2016/04/28/google
-ceo-cites-shift-from-a-mobile-world-to-an-ai-world.html.

18. Apple from the Top Rope

1 Adi Robertson, "Apple Is Launching an iOS 'ARKit' for Augmented Reality Apps,"
The Verge, June 5, 2017, https://web.archive.org/web/20170605203417/https://
www.theverge.com/2017/6/5/15732832/apple-augmented-reality-arkit-ar-sdk
-wwdc-2017.

2 Stephen Cosman et al., "User Behavior Model Development with Private
Federated Learning," United States US20210192078A1, filed December 21, 2020,
and issued July 30, 2024, https://patents.google.com/patent/US20210192078A1
/en; Abhishek Bhowmick et al., "Protection against Reconstruction and Its
Applications in Private Federated Learning" (arXiv, June 3, 2019), https://
doi.org/10.48550/arXiv.1812.00984; Amit Naik, "For the Sake of Privacy:
Apple's Federated Learning Approach," *Analytics India Magazine*, November 3,
2021, https://analyticsindiamag.com/ai-features/for-the-sake-of-privacy-apples
-federated-learning-approach/; Matthias Paulik et al., "Federated Evaluation and
Tuning for On-Device Personalization: System Design and Applications" (arXiv,
February 16, 2021), https://doi.org/10.48550/arXiv.2102.08503.

3 H. Brendan McMahan et al., "Communication-Efficient Learning of Deep
Networks from Decentralized Data" (arXiv, January 26, 2023), https://doi
.org/10.48550/arXiv.1602.05629; Naik, "For the Sake of Privacy."

4 Brendan McMahan and Daniel Ramage, "Federated Learning: Collaborative
Machine Learning without Centralized Training," *Google Research* (blog), April 6,

2017, https://research.google/blog/federated-learning-collaborative-machine-learning-without-centralized-training-data/.

5　David Murakami Wood and David Eliot, "Minding the FLoCs: Google's Marketing Moves, AI, Privacy and the Data Commons," Centre for International Governance Innovation, May 20, 2021, https://www.cigionline.org/articles/minding-flocs-googles-marketing-moves-ai-privacy-and-data-commons/.

6　Bhowmick et al., "Protection against Reconstruction"; Paulik et al., "Federated Evaluation."

7　Paulik et al., "Federated Evaluation."

8　"Apple Unleashes M1," Apple Newsroom (Canada), November 10, 2020, https://www.apple.com/ca/newsroom/2020/11/apple-unleashes-m1/; Prince Onyeanuna, "Overview of the Apple M1 Chip Architecture," EverythingDevOps, July 24, 2024, https://everythingdevops.dev/overview-of-the-apple-m1-chip-architecture/.

9　Nermin Hajdarbegović, "Apple M1 Overview and Compatibility," *Toptal Engineering Blog* (blog), https://www.toptal.com/ios/apple-m1-processor-compatibility-overview.

10　Dan Frommer, "Apple's AR Glasses Project Has a Leader and a Team," *Vox*, March 20, 2017, https://www.vox.com/2017/3/20/14981470/apple-ar-glasses-mike-rockwell.

11　Nick Bilton, "Why Tim Cook Is Going All In on the Apple Vision Pro," *Vanity Fair*, February 1, 2024, https://www.vanityfair.com/news/tim-cook-apple-vision-pro.

12　Tim Bradshaw and Patrick McGee, "Tim Cook Bets on Apple's Mixed-Reality Headset to Secure His Legacy," *Financial Times*, March 12, 2023, sec. Apple Inc, https://www.ft.com/content/86b99549-0648-4c23-ab6e-642a4ba51dff.

13　"Introducing Apple Vision Pro: Apple's First Spatial Computer," Apple Newsroom (Canada), press release, June 5, 2023, https://www.apple.com/ca/newsroom/2023/06/introducing-apple-vision-pro/.

14　Yildiz et al., "Creation of Optimal Working, Learning, and Resting Environments on Electronic Devices," United States US20210096646A1, filed September 24, 2020, and issued October 10, 2023, https://patents.google.com/patent/US20210096646A1/en.

15　"Introducing Apple Vision Pro."

16　Nilay Patel, "Apple Vision Pro Review: Magic, Until It's Not," *The Verge*, January 30, 2024, https://www.theverge.com/24054862/apple-vision-pro-review-vr-ar-headset-features-price; Qianer Liu, Patrick McGee, and Kana Inagaki, "Apple Forced to Make Major Cuts to Vision Pro Headset Production Plans," *Financial Times*, July 3, 2023, sec. Apple Inc, https://www.ft.com/content/b6f06bde-17b0-4886-b465-b561212c96a9; Mark Sullivan, "With the Vision Pro, Apple Has Never Depended More on Developers for a Product's Success," Fast Company, January 26, 2024, https://www.fastcompany.com/91017225/apple-vision-pro-developers-sales.

17　"Apple's 2025 Plan: iPhone Overhaul, Smart Home Push and AI Catch-Up," *Bloomberg*, January 12, 2025, https://www.bloomberg.com/news/newsletters/2025-01-12/apple-2025-plans-iphone-17-smart-home-hub-ios-19-ai-apple-watch-ipads-m5; Matthew Phelan, "Apple Quietly Discontinuing Flagship Device Due to Lackluster Sales," *Daily Mail*, November 11, 2024, sec. Science, https://www.dailymail.co.uk/sciencetech/article-14068925/Apple-quietly-discontinuing-flagship-device-lackluster-sales.html.

18 "Ray-Ban Meta AI Glasses," Meta, https://www.meta.com/ca/ai-glasses/.

19 Alex Heath, "This Is Meta's AR/VR Hardware Roadmap through 2027," *The Verge*, February 28, 2023, https://www.theverge.com/2023/2/28/23619730/meta-vr-oculus -ar-glasses-smartwatch-plans.

20 "Introducing Orion, Our First True Augmented Reality Glasses," *Meta* (blog), September 25, 2024, https://about.fb.com/news/2024/09/introducing-orion -our-first-true-augmented-reality-glasses/.

21 "Brilliant Labs," https://brilliant.xyz/.

22 Paresh Dave, "Humane's AI Pin Is a $700 Smartphone Alternative You Wear All Day," *Wired*, November 9, 2023, https://www.wired.com/story/humane-ai -pin-700-dollar-smartphone-alternative-wearable/; David Pierce, "Humane AI Pin Review: Not Even Close," *The Verge*, April 11, 2024, https://www.theverge .com/24126502/humane-ai-pin-review.

23 "The World's Most Advanced Control Program," Naqi Logix, https://www.naqi- logix.com.

19. The Surveillance State

1 For further reading on the Stasi, see John O. Koehler, *Stasi: The Untold Story of the East German Secret Police* (Perseus Books, 1999); Jens Gieseke, *The History of the Stasi: East Germany's Secret Police, 1945–1990* (Berghahn Books, 2014); Anna Funder, *Stasiland: Stories from behind the Berlin Wall* (Harper Collins, 2011); Charles River, *The Stasi: The History and Legacy of East Germany's Secret Police Agency* (CreateSpace Independent Publishing Platform, 2018).

2 "Stasi Records Archive," CIPDH International Center for the Promotion of Human Rights–UNESCO, https://www.cipdh.gob.ar/memorias-situadas/en/ lugar-de-memoria/archivos-de-la-stasi/.

3 Joel D. Cameron, "Stasi: East German Government," in *Encyclopedia Britannica*, February 24, 2025, https://www.britannica.com/topic/Stasi.

4 Martijn van Otterlo, "Automated Experimentation in Walden 3.0.: The Next Step in Profiling, Predicting, Control and Surveillance," *Surveillance & Society* 12, no. 2 (May 9, 2014): 255–72, https://doi.org/10.24908/ss.v12i2.4600.

5 David Lyon, *The Electronic Eye: The Rise of Surveillance Society*, new edition (University of Minnesota Press, 1994), ix, https://www.jstor.org/stable/10.5749/j .ctttsqw8.

6 Michael Hatfield, "Taxation and Surveillance: An Agenda," SSRN Scholarly Paper (Social Science Research Network, December 17, 2014), https://doi .org/10.2139/ssrn.2539835; Anjana Haines, "Tax Watch: The Decade of Digital Surveillance," *International Tax Review*, February 17, 2020, https://www.proquest .com/openview/05a5de1635159084199c35508c1028b4/1; Jordan Burrows, "IRS Using Artificial Intelligence to Make Sure People Aren't Playing the System," *CBS News*, March 27, 2024, https://www.cbsnews.com/detroit/news/irs-using -artificial-intelligence-ai-taxes/.

7 For further discussion about legibility and state-led social planning as it relates to taxes and government surveillance more broadly, see James C. Scott, *Seeing Like a State: How Certain Schemes to Improve the Human Condition Have Failed* (Yale University Press, 1998).

8 Scott, *Seeing Like a State*, 65.

9 Scott, 65–6.

10 Scott, 66–7.

11 Scott, 68.

12 Scott, *Seeing Like a State*, 69; Francis Alvarez Gealogo, "Looking for Claveria's Children: Church, State, Power, and the Individual in Philippine Naming Systems during the Late Nineteenth Century," in *Personal Names in Asia: History, Culture and Identity*, by Zheng Yangwen and Charles J-H Macdonald (NUS Press, 2010), 37–51, https://www.jstor.org/stable/j.ctv1qv2k0.7.

13 James B. Rule et al., "Documentary Identification and Mass Surveillance in the United States," *Social Problems* 31, no. 2 (December 1983): 229, https://doi .org/10.2307/800214.

14 Kirstie Ball et al., *A Report on the Surveillance Society: For the Information Commissioner by the Surveillance Studies Network*, September 2006, 13, https://ico.org.uk/media /about-the-ico/documents/1042390/surveillance-society-full-report-2006.pdf.

15 Shoshana Zuboff, *The Age of Surveillance Capitalism: The Fight for a Human Future at the New Frontier of Power* (PublicAffairs, 2019).

16 Josh Lauer, "Coming to Terms with Credit: The Nineteenth-Century Origins of Consumer Credit Surveillance," in *Creditworthy: A History of Consumer Surveillance and Financial Identity in America* (Columbia University Press, 2017), 51–77, http:// www.jstor.org/stable/10.7312/laue16808.6; David Lyon, "Introduction: Body, Soul and Credit Card," in *The Electronic Eye: The Rise of Surveillance Society* (University of Minnesota Press, 1994), 3–21, http://www.jstor.org/stable/10.5749/j.ctttsqw8.4.

17 Brett Williams, *Debt for Sale: A Social History of the Credit Trap* (University of Pennsylvania Press, 2011), 33–61, https://doi.org/10.9783/9780812200782; Lawrence M. Ausubel, "The Failure of Competition in the Credit Card Market," *The American Economic Review* 81, no. 1 (1991): 50–81, https://www.jstor.org /stable/2006788.

18 Syntropy Group, "How to Use Credit Bureau Data for Targeted Marketing," February 16, 2023, https://syntropygroup.com/how-to-use-credit-bureau-data-for -targeted-marketing/.

19 David Murakami Wood and Kirstie Ball, "Brandscapes of Control? Surveillance, Marketing and the Co-construction of Subjectivity and Space in Neo-Liberal Capitalism," *Marketing Theory* 13, no. 1 (2013): 47–67, https://doi .org/10.1177/1470593112467264; Joseph Turow and Nora Draper, "Advertising's New Surveillance Ecosystem," in *Routledge Handbook of Surveillance Studies*, ed. Kirstie Ball, Kevin D. Haggerty, and David Lyon (Routledge, 2012), 133–40, https://doi.org/10.4324/9780203814949.

20 Richard James Webber, "The Evolution of Direct, Data and Digital Marketing," *Journal of Direct Data and Digital Marketing Practice* 14, no. 4 (April 2013): 294 –5, https://doi.org/10.1057/dddmp.2013.20; Thomas Smith, "The History of Marketing Tools, Their Big Moments, and Their Future" (thesis, Murray State University, 2019), Murray State's Digital Commons, https://digitalcommons .murraystate.edu/bis437/199.

21 Shoshana Zuboff, "Big Other: Surveillance Capitalism and the Prospects of an Information Civilization," *Journal of Information Technology* 30, no. 1 (2015): 75–89, https://doi.org/10.1057/jit.2015.5; Christian Fuchs, "Google Capitalism," *Triple C* 10, no. 1 (2012): 42–8, https://doi.org/10.31269/triplec.v10i1.304.

22 David Eliot and David Murakami Wood, "Culling the FLoC: Market Forces, Regulatory Regimes and Google's (Mis)Steps on the Path Away from Targeted Advertising," *Information Polity* 27, no. 2 (2022): 259, https://doi.org/10.3233/IP-211535.

23 Alex Krizhevsky, Ilya Sutskever, and Geoffrey E. Hinton, "ImageNet Classification with Deep Convolutional Neural Networks," in *Advances in Neural Information*

Processing Systems, vol. 25, ed. F. Pereira et al. (Curran Associates, 2012), https://proceedings.neurips.cc/paper_files/paper/2012/hash/c399862d3b9d6b76c8436e924a68c45b-Abstract.html.

24 Emily Denton et al., "On the Genealogy of Machine Learning Datasets: A Critical History of ImageNet," *Big Data & Society* 8, no. 2 (2021), https://doi.org/10.1177/20539517211035955.

25 Melany Amarikwa, "Internet Openness at Risk: Generative AI's Impact on Data Scraping," *Richmond Journal of Law & Technology* 30, no. 3 (2024): 541, 543, 565, https://doi.org/10.2139/ssrn.4723713; for further reading on AI and web scraping, see Manushi Weerasinghe, "Enhancing Web Scraping with Artificial Intelligence: A Review" (paper presented at the 4th Research Symposium of Faculty of Computing 2024, General Sir John Kotelawala Defence University, January 17, 2024), https://www.researchgate.net/publication/379024314_Enhancing_Web_Scraping_with_Artificial_Intelligence_A_Review.

26 Mark Andrejevic, "Surveillance in the Digital Enclosure," *The Communication Review* 10, no. 4 (2007): 295–317, https://doi.org/10.1080/10714420701715365; Mark Andrejevic, "Meta-Surveillance in the Digital Enclosure," *Surveillance & Society* 20, no. 4 (2022): 392, https://doi.org/10.24908/ss.v20i4.16008.

27 Andrejevic, "Surveillance in the Digital Enclosure"; Andrejevic, "Meta-Surveillance in the Digital Enclosure," 392.

28 Andrejevic, "Meta-Surveillance in the Digital Enclosure," 391.

29 "Regulation (EU) 2016/679 of the European Parliament and of the Council," *Official Journal of the European Union* 119, no. 4.5 (April 27, 2016): 1–88, https://eur-lex.europa.eu/legal-content/EN/TXT/PDF/?uri=CELEX:32016R0679; "General Data Protection Regulation (GDPR)," Intersoft Consulting, https://gdpr-info.eu/.

20. Data and Power

1 "The World's Most Valuable Resource Is No Longer Oil, but Data," *The Economist*, May 6, 2017, sec. Leaders, https://www.economist.com/leaders/2017/05/06/the-worlds-most-valuable-resource-is-no-longer-oil-but-data; Kiran Bhageshpur, "Council Post: Data Is the New Oil – And That's a Good Thing," *Forbes*, November 15, 2019, https://www.forbes.com/councils/forbestechcouncil/2019/11/15/data-is-the-new-oil-and-thats-a-good-thing/.

2 Nadya Purtova and Gijs van Maanen, "Data as an Economic Good, Data as a Commons, and Data Governance," *Law, Innovation and Technology* 16, no. 1 (2024): 11, https://doi.org/10.1080/17579961.2023.2265270.

3 James Hall, Laura Glitsos, and Jess Taylor, "Fungible," *M/C Journal* 25, no. 2 (2022), https://doi.org/10.5204/mcj.2905.

4 Purtova and van Maanen, "Data as an Economic Good," 16; Roxana Mihet and Thomas Philippon, "The Economics of Big Data and Artificial Intelligence," in *Disruptive Innovation in Business and Finance in the Digital World*, ed. J. Jay Choi and Bora Ozkan, *International Finance Review*, vol. 20 (Emerald Publishing, 2019), 30, https://doi.org/10.1108/S1569-376720190000020006.

5 Jia Deng et al., "ImageNet: A Large-Scale Hierarchical Image Database," in *2009 IEEE Conference on Computer Vision and Pattern Recognition* (IEEE, 2009), 248–55, https://doi.org/10.1109/CVPR.2009.5206848; Emily Denton et al., "On the Genealogy of Machine Learning Datasets: A Critical History of ImageNet," *Big Data & Society* 8, no. 2 (July 1, 2021), https://doi.org/10.1177/20539517211035955.

6 Lisa Gitelman, ed., *"Raw Data" Is an Oxymoron* (MIT Press, 2013), 3.
7 For further reading, see Patrik Hummel, Matthias Braun, and Peter Dabrock, "Own Data? Ethical Reflections on Data Ownership," *Philosophy & Technology* 34, no. 3 (September 1, 2021): 545–72, https://doi.org/10.1007/s13347-020-00404 -9; Jacqueline Hicks, "The Future of Data Ownership: An Uncommon Research Agenda," *The Sociological Review* 71, no. 3 (May 1, 2023): 544–60, https://doi .org/10.1177/00380261221088120; Shilei Li, Yang Liu, and Juan Feng, "Who Should Own the Data? The Impact of Data Ownership Shift from the Service Provider to Consumers," *Journal of Management Information Systems* 40, no. 2 (April 3, 2023): 366–400, https://doi.org/10.1080/07421222.2023.2196775; Teresa Scassa, "Data Ownership," Centre for International Governance Innovation, CIGI Papers, 187 (September 2018), https://www.cigionline.org/static/documents /documents/Paper%20no.187_2.pdf; Ignacio Cofone, "Beyond Data Ownership," SSRN Scholarly Paper (Social Science Research Network, March 30, 2020), https://doi.org/10.2139/ssrn.3564480.
8 Scassa, "Data Ownership," 14.
9 "Your Data Is Shared and Sold ... What's Being Done about It?," *Knowledge at Wharton*, October 28, 2019, https://knowledge.wharton.upenn.edu/article/data -shared-sold-whats-done/.
10 "Regulation (EU) 2016/679 of the European Parliament and of the Council of 27 April 2016 on the protection of natural persons with regard to the processing of personal data and on the free movement of such data, and repealing Directive 95/46/EC (General Data Protection Regulation)," *Official Journal of the European Union*, L119, May 4, 2016, 1–88, art. 15, https://eur-lex.europa.eu/legal-content /EN/TXT/PDF/?uri=CELEX:32016R0679.
11 Shoshana Zuboff, "Big Other: Surveillance Capitalism and the Prospects of an Information Civilization," *Journal of Information Technology* 30 (April 4, 2015): 81, https://doi.org/10.1057/jit.2015.5.
12 Nick Srnicek, *Platform Capitalism* (Polity Press, 2017).
13 Brett Aho, "Data Communism: Constructing a National Data Ecosystem," *Big Data & Society* 11, no. 3 (2024), https://doi.org/10.1177/20539517241275888.
14 See "Opinions of the Central Committee of the Communist Party of China and the State Council on Building a More Complete System and Mechanism for Market-Based Allocation of Factors" (State Council, 2020), as referenced in Aho, "Data Communism."
15 "Opinions of the Central Committee."
16 Bianca Wylie and Sean Martin McDonald, "What Is a Data Trust?," Centre for International Governance Innovation, October 9, 2018, https://www.cigionline .org/articles/what-data-trust/; Digital Public, "Digital Content Governance and Data Trusts – Diversity of Content in the Digital Age," Government of Canada, February 2020, https://www.canada.ca/en/canadian-heritage/services/diversity -content-digital-age/digital-content-governance-data-trust.html; "About Digital Trust," World Economic Forum, https://initiatives.weforum.org/digital-trust /home; Anouk Ruhaak et al., "A Practical Framework for Applying Ostrom's Principles to Data Commons Governance," Mozilla Foundation, December 6, 2021, https://foundation.mozilla.org/en/blog/a-practical-framework-for -applying-ostroms-principles-to-data-commons-governance/.
17 Michael Haupt, "Introducing Personal Data Exchanges & the Personal Data Economy," *Medium* (blog), December 7, 2016, https://medium.com /project-2030/what-is-a-personal-data-exchange-256bcd5bf447.

21. The Ubiquitous Machine

1 A philosophical position widely popularized by Langdon Winner. See Langdon Winner, "Do Artifacts Have Politics?" *Daedalus* 109, no. 1 (Winter 1980): 121–36, https://www.jstor.org/stable/20024652.

2 I wrote this part as a hypothetical in March 2024. However, while doing the final readthrough of the book in January 2025, I Googled "pizza places near me," and Pizza Hut, with a "Sponsored" tag, was in fact the first search result.

3 Sophie Bishop, "Algorithmic Experts: Selling Algorithmic Lore on YouTube," *Social Media + Society* 6, no. 1 (2020), https://doi.org/10.1177/2056305119897323.

4 Bishop, "Algorithmic Experts"; Sophie Bishop, "Anxiety, Panic and Self-Optimization: Inequalities and the YouTube Algorithm," *Convergence* 24, no. 1 (2018): 69–84, https://doi.org/10.1177/1354856517736978; Tim Peterson, "Creators Are Making Longer Videos to Cater to the YouTube Algorithm," *Digiday*, July 3, 2018, https://digiday.com/future-of-tv/creators-making-longer-videos-cater-youtube-algorithm/; Hana Kiros, "Hated That Video? YouTube's Algorithm Might Push You Another Just Like It," *MIT Technology Review*, September 20, 2022, https://www.technologyreview.com/2022/09/20/1059709/youtube-algorithm-recommendations/.

5 Renkai Ma and Yubo Kou, "'How Advertiser-Friendly Is My Video?': YouTuber's Socioeconomic Interactions with Algorithmic Content Moderation," *Proceedings of the ACM on Human-Computer Interaction* 5, no. CSCW2 (2021): 1–25, https://doi.org/10.1145/3479573.

6 "How Does the YouTube Algorithm Work: A 2024 Guide," Influencer Marketing Hub, September 12, 2024, https://influencermarketinghub.com/how-does-the-youtube-algorithm-work/.

7 Ilias Papastratis et al., "AI Nutrition Recommendation Using a Deep Generative Model and ChatGPT," *Scientific Reports* 14, no. 1 (June 25, 2024): 14620, https://doi.org/10.1038/s41598-024-65438-x.

8 Jose Sanchez Gracias et al., "Smart Cities – A Structured Literature Review," *Smart Cities* 6, no. 4 (August 2023): 1719–43, https://doi.org/10.3390/smartcities6040080.

9 Chuck Brooks, "The Emergence of Smart Cities in the Digital Era," *Forbes*, November 18, 2023, https://www.forbes.com/sites/chuckbrooks/2023/11/18/the-emergence-of-smart-cities-in-the-digital-era/.

10 Maria-Alexandra Barina and Gabriel Barina, "From Elusive to Ubiquitous: Understanding Smart Cities" (arXiv, April 24, 2020), https://doi.org/10.1145/3479573; S. Aslam and H. Sami Ullah, "A Comprehensive Review of Smart Cities Components, Applications, and Technologies Based on Internet of Things" (arXiv, February 5, 2020), https://doi.org/10.48550/arXiv.2002.01716.

11 Monique Mann et al., "#BlockSidewalk to Barcelona: Technological Sovereignty and the Social License to Operate Smart Cities," *Journal of the Association for Information Science and Technology* 71, no. 9 (2020): 1103–15, https://doi.org/10.1002/asi.24387.

12 Mann et al., "#BlockSidewalk," 9–10.

13 Josep-Ramon Ferrer, "Barcelona's Smart City Vision: An Opportunity for Transformation," *Field Actions Science Reports: The Journal of Field Actions*, Special Issue 16 (June 1, 2017): 70–5, https://journals.openedition.org/factsreports/4367; Ignasi Capdevila and Matías I. Zarlenga, "Smart City or Smart Citizens? The

Barcelona Case," *Journal of Strategy and Management* 8, no. 3 (August 2015): 266–82, https://doi.org/10.1108/JSMA-03-2015-0030.

14 "Smart Citizen: Open Tools for Environmental Monitoring," Fab Lab Barcelona, March 19, 2020, https://fablabbcn.org/projects/smart-citizen.

15 Laura Adler, "How Smart City Barcelona Brought the Internet of Things to Life," Data-Smart City Solutions, February 18, 2016, https://datasmart.hks.harvard .edu/news/article/how-smart-city-barcelona-brought-the-internet-of-things-to -life-789.

16 "Barcelona: Nature-Based Solutions (NbS) Enhancing Resilience to Climate Change," Oppla, January 26, 2017, https://oppla.eu/casestudy/17283; Lydia Chaparro and James Terradas, *Ecological Services of Urban Forest in Barcelona* (Universitat Autònoma de Barcelona-CREAF; Ajuntament de Barcelona, 2009), https://www.itreetools.org/documents/302/Barcelona%20Ecosystem%20 Analysis.pdf; "Tree Master Plan," Urban Nature Atlas, November 2021, https:// una.city/nbs/barcelona/tree-master-plan.

17 Ronika Postaria, "Superblock (Superilla) Barcelona – A City Redefined," *Cities Forum* (blog), May 31, 2021, https://www.citiesforum.org/news/superblock -superilla-barcelona-a-city-redefined/; European Bank for Reconstruction and Development (EBRD), "Urban Planning with Superblocks: Barcelona, Spain," *EBRD Green Cities*, https://www.ebrdgreencities.com/policy-tool/urban-planning -with-superblocks-barcelona-spain-2/.

18 Adler, "How Smart City Barcelona Brought the Internet of Things to Life."

22. The Apple

1 David Leavitt, *The Man Who Knew Too Much: Alan Turing and the Invention of the Computer* (W.W. Norton, 2006), 266–8; Walter Isaacson, *The Innovators: How a Group of Hackers, Geniuses, and Geeks Created the Digital Revolution* (Simon and Schuster, 2014), 129.

2 Andrew Hodges, *Alan Turing: The Enigma* (Random House, 2012), 33–4, Apple Books.

3 Hodges, *Alan Turing: The Enigma*, 1098; Leavitt, *The Man Who Knew Too Much*, 4.

4 Isaacson, *The Innovators*, 129.

5 Isaacson, *The Innovators*, 129; Leavitt, *The Man Who Knew Too Much*, 268.

6 It should be noted that, although Dr. Turing taking his life with a cyanide apple is the "official" story, there is historical debate regarding its truth. There is justified skepticism regarding Turing's cause of death and speculation that it may have been an accident and not a suicide. If it was a suicide, it is unclear if a cyanide apple was truly the cause of death or a myth born out of non-evidence-driven assumptions made by the investigators. See B. Jack Copeland et al., *The Turing Guide* (Oxford University Press, 2017), 14–17; Hodges, *Alan Turing: The Enigma*, 1162; Leavitt, *The Man Who Knew Too Much*, 268–80; B. Jack Copeland, *Turing: Pioneer of the Information Age* (Oxford University Press, 2014).

7 Edwin Black, *IBM and the Holocaust: The Strategic Alliance between Nazi Germany and America's Most Powerful Corporation*, expanded ed. (Dialog Press, 2001).

8 Jesse Dillard, "Professional Services, IBM, and the Holocaust," *Journal of Information Systems* 17, no. 2 (September 2003): 3, https://doi.org/10.2308/jis.2003.17.2.1; Zygmunt Bauman, *Modernity and the Holocaust* (Cornell University Press, 2002), 14–17.

9 Black, *IBM and the Holocaust*, 52–4.
10 Black, 55.
11 James W. Cortada, "Change and Continuity at IBM: Key Themes in Histories of IBM," *Business History Review* 92, no. 1 (April 2018): 117–48, https://doi .org/10.1017/S0007680518000041.
12 Cortada, "Change and Continuity"; Dillard, "Professional Services," 8.
13 Dillard, "Professional Services," 5.
14 Harry Murphy, "Dealing with the Devil: The Triumph and Tragedy of IBM's Business with the Third Reich," *The History Teacher* 53, no. 1 (2019): 176, https:// www.jstor.org/stable/27058571.
15 Edwin Black, "The Nazi Party: IBM & 'Death's Calculator,'" Jewish Virtual Library, https://www.jewishvirtuallibrary.org/ibm-and-quot-death-s-calculator-quot-2.
16 Black, *IBM and the Holocaust*, 54–5, 199.
17 Edwin Black, "IBM's Role in the Holocaust – What the New Documents Reveal," *HuffPost*, February 27, 2012, https://www.huffpost.com/entry/ibm -holocaust_b_1301691.
18 Donna K. Nagata, Jacqueline H.J. Kim, and Kaidi Wu, "The Japanese American Wartime Incarceration: Examining the Scope of Racial Trauma," *The American Psychologist* 74, no. 1 (January 2019): 36–48, https://doi.org/10.1037 /amp0000303; "Japanese-American Incarceration during World War II," National Archives, August 15, 2016, https://www.archives.gov/education /lessons/japanese-relocation; "Terminology and the Mass Incarceration of Japanese Americans during World War II," US National Park Service, https:// www.nps.gov/articles/000/terminology-and-the-mass-incarceration-of-japanese -americans-during-world-war-ii.htm; Mayya Komisarchik, Maya Sen, and Yamil R. Velez, "The Political Consequences of Ethnically Targeted Incarceration: Evidence from Japanese-American Internment during WWII," *The Journal of Politics* 84, no. 3 (2022): 1497–514, https://doi.org/10.1086/717262.
19 Thomas Tyson and Richard Fleischman, "Accounting for Interned Japanese-American Civilians during World War II: Creating Incentives and Establishing Controls for Captive Workers," *Accounting Historians Journal* 33, no. 1 (January 1, 2006): 167, https://egrove.olemiss.edu/aah_journal/vol33/iss1/19.
20 Associate Justice Owen J. Roberts, in his dissenting opinion in the *Korematsu* ruling (about the constitutionality of the camps), used the term "concentration camp." The relocation centers have since been recognized as "concentration camps" in post-war and dissenting accounts, including the 1980s Commission on Wartime Relocation and Internment of Civilians (CWRIC), which was commissioned by the US Congress to study the internment.
21 "Press Conference 982, November 21, 1944," Franklin D. Roosevelt Library (Hyde Park, New York), 246, cited by Roger Daniels, "Words Do Matter: A Note on Inappropriate Terminology and the Incarceration of the Japanese Americans," in *Nikkei in the Pacific Northwest: Japanese Americans and Japanese Canadians in the Twentieth Century*, ed. Louis Fiset and Gail Nomura (University of Washington Press, 2005), 183–207, https://npshistory.com/publications/incarceration /words-do-matter.pdf.
22 "Japanese-American Incarceration during World War II," National Archives.
23 Excerpts from Jeffery F. Burton et al., *Confinement and Ethnicity: An Overview of World War II Japanese American Relocation Sites* (University of Washington Press, 2011), summarized in "A Brief History of Japanese American Relocation during World War II," US National Park Service, https://www.nps.gov/articles /historyinternment.htm.

24 "Japanese-American Incarceration during World War II," National Archives; "Japanese American Incarceration," The National WWII Museum, New Orleans, https://www.nationalww2museum.org/war/articles/japanese-american-incarceration.

25 "Japanese-American Incarceration during World War II," National Archives; "Japanese American Incarceration," The National WWII Museum, New Orleans.

26 Tyson and Fleischman, "Accounting for Interned Japanese-American Civilians," 189–90.

27 Tyson and Fleischman, 191.

28 Wendy Ng, *Japanese American Internment during World War II: A History and Reference Guide* (Greenwood Press, 2002), 57; Brian Masaru Hayashi, *Democratizing the Enemy: The Japanese American Internment* (Princeton University Press, 2004), 1–2.

29 Ng, *Japanese American Internment*, xxii, 38, 57–9.

30 Natasha Varner, "The 'Loyalty Questionnaire' of 1943 Opened a Wound That Has Yet to Heal," Densho, July 19, 2019, https://densho.org/catalyst/the-loyalty-questionnaire-of-1943-opened-a-wound-that-has-yet-to-heal/.

31 "Personal Justice Denied: Report of the Commission on Wartime Relocation and Internment of Civilians," National Archives, https://www.archives.gov/research/aapi/ww2/justice.

32 Nagata, Kim, and Wu, "The Japanese American Wartime Incarceration," 36.

33 Nagata, Kim, and Wu, 40.

34 Deborah Ellis, "'The Arc of the Moral Universe Is Long, but It Bends toward Justice,'" *Obama White House Archives* (blog), October 21, 2011, https://obamawhitehouse.archives.gov/blog/2011/10/21/arc-moral-universe-long-it-bends-toward-justice.

35 Kate Ellis and Stephen Smith, "The FBI's War on King," *ARMreports*, sec. King's Last March, https://features.apmreports.org/arw/king/d1.html.

36 Jeffrey Haas, *The Assassination of Fred Hampton: How the FBI and the Chicago Police Murdered a Black Panther* (Chicago Review Press, 2019).

37 Sam Biddle and Maryam Saleh, "Little-Known Federal Software Can Trigger Revocation of Citizenship," *The Intercept*, August 25, 2021, https://theintercept.com/2021/08/25/atlas-citizenship-denaturalization-homeland-security/.

38 Biddle and Saleh, "Little-Known Federal Software."

39 Biddle and Saleh, "Little-Known Federal Software"; Access Now et al. to Lynn Parker Dupree and Amanda Baran, "Follow Up Letter to DHS Privacy & USCIS," May 31, 2022, Government Information Watch, https://www.govinfowatch.net/wp-content/uploads/2023/02/Follow-Up-Letter-DHS-Privacy-and-USCIS-5.31.22_FINAL.pdf.

40 Biddle and Saleh, "Little-Known Federal Software."

41 "Privacy Impact Assessment for the ATLAS," Department of Homeland Security, U.S. Citizenship and Immigration Services, October 30, 2020, https://www.dhs.gov/sites/default/files/publications/privacy-pia-uscis084-atlas-july2021.pdf; Access Now et al. to Lynn Parker Dupree and Amanda Baran, "Follow Up Letter," 4.

42 Kate Crawford, *The Atlas of AI: Power, Politics, and the Planetary Costs of Artificial Intelligence* (Yale University Press, 2021); Kate Crawford, "Can an Algorithm Be Agonistic? Ten Scenes from Life in Calculated Publics," *Science, Technology, & Human Values* 41, no. 1 (2016): 77–92; Mike Ananny and Kate Crawford, "Seeing without Knowing: Limitations of the Transparency Ideal and Its Application to Algorithmic Accountability," *New Media & Society* 20, no. 3 (2018): 973–89, https://doi.org/10.1177/1461444816676645; Cathy O'Neil, *Weapons of Math Destruction: How Big Data Increases Inequality and Threatens Democracy* (Crown Publishing Group, 2016); Cathy O'Neil, *The Shame Machine: Who Profits in the New*

Age of Humiliation (Crown Publishing Group, 2022); Elizabeth E. Joh, "The Undue Influence of Surveillance Technology Companies on Policing," *New York University Law Review* 92, no. 19 (2017): 19–47, https://doi.org/10.2139/ssrn.2924620; Elizabeth E. Joh, "The New Surveillance Discretion: Automated Suspicion, Big Data, and Policing," *Harvard Law & Policy Review* 10 (2016): 15–42, https://journals.law.harvard.edu/lpr/wp-content/uploads/sites/89/2016/02/10.1_3_Joh.pdf; Elizabeth E. Joh, "Police Technology Experiments," *Columbia Law Review Forum* 125, no. 1 (January 31, 2025): 1–28, https://doi.org/10.2139/ssrn.4721955; Elizabeth Joh, "Ethical AI in American Policing," *Notre Dame Journal on Emerging Technologies* 3, no. 2 (November 2022): 261–87, https://scholarship.law.nd.edu/ndlsjet/vol3/iss2/4; Elizabeth Joh, "Policing by Numbers: Big Data and the Fourth Amendment," *Washington Law Review* 89, no. 35 (March 1, 2014): 35–68, https://digitalcommons.law.uw.edu/wlr/vol89/iss1/3; Sarah Brayne, "Big Data Surveillance: The Case of Policing," *American Sociological Review* 82, no. 5 (October 1, 2017): 977–1008, https://doi.org/10.1177/0003122417725865; Sarah Brayne, "Surveillance and System Avoidance: Criminal Justice Contact and Institutional Attachment," *American Sociological Review* 79, no. 3 (2014): 367–91, https://doi.org/10.1177/0003122414530398; Sarah Brayne, "The Banality of Surveillance," *Surveillance & Society* 20, no. 4 (December 16, 2022): 372–8, https://doi.org/10.24908/ss.v20i4.15946; Timnit Gebru, "Race and Gender," in *The Oxford Handbook of Ethics of AI*, ed. Markus D. Dubber, Frank Pasquale, and Sunit Das (Oxford University Press, 2020), 252–69, https://doi.org/10.1093/oxfordhb/9780190067397.013.16.

43 The field of computational fluid dynamics (CFD) has progressed substantially in recent years; however, it is still limited in application. The major issue is that turbulent (in contrast to laminar) flows are extremely chaotic: turbulent flows contain intricate multiscale structures that continuously form, interact, and dissipate. The core difficulty with CFD and turbulence modeling lies in the need for extremely fine spatial and temporal resolution to capture all scales of motion, leading to high computational costs. Oddly, my favorite painter, Van Gogh, was extremely adept at modeling the turbulent flows of air (without a computer!) in his masterpiece *The Starry Night*. See Yinxiang (马寅翔) Ma et al., "Hidden Turbulence in Van Gogh's *The Starry Night*," *Physics of Fluids* 36, no. 9 (September 17, 2024): 095140, https://doi.org/10.1063/5.0213627.

44 Lisa Gitelman, ed., *"Raw Data" Is an Oxymoron* (MIT Press, 2013), 3.

45 Hope Reese, "What Happens When Police Use AI to Predict and Prevent Crime?," *JSTOR Daily*, February 23, 2022, https://daily.jstor.org/what-happens-when-police-use-ai-to-predict-and-prevent-crime/; EUCPN and Majsa Storbeck, *Artificial Intelligence and Predictive Policing: Risks and Challenges* (EUCPN, 2022), https://eucpn.org/sites/default/files/document/files/PP%20%282%29.pdf.

46 Brayne, "Big Data Surveillance."

47 Brayne, "Big Data Surveillance"; Anthony Allan Braga and David Weisburd, *Policing Problem Places: Crime Hot Spots and Effective Prevention* (Oxford University Press, 2010); Sarah Brayne, Alex Rosenblat, and Danah Boyd, "Predictive Policing" (paper presented at the conference Data & Civil Rights: A New Era of Policing and Justice, Washington, DC, October 27, 2015), 7, https://datacivilrights.org/pubs/2015-1027/Predictive_Policing.pdf.

48 Brayne, "Big Data Surveillance"; Braga and Weisburd, *Policing Problem Places*; Brayne, Rosenblat, and Boyd, "Predictive Policing"; Andrew Guthrie Ferguson,

The Rise of Big Data Policing: Surveillance, Race, and the Future of Law Enforcement (NYU Press, 2017), 34–61; Joh, "The New Surveillance Discretion."

49 EUCPN and Storbeck, *Artificial Intelligence and Predictive Policing.*

50 Takara Small, "The Downside of AI: Former Google Scientist Timnit Gebru Warns of the Technology's Built-in Biases," *Globe and Mail*, May 12, 2023, https:// www.theglobeandmail.com/business/article-the-downside-of-ai-former-google -scientist-timnit-gebru-warns-of-the/.

51 Simone Browne, *Dark Matters: On the Surveillance of Blackness* (Duke University Press, 2015), 111; Samantha Cole, "Amazon Pulled the Plug on an AI Recruitment Tool That Was Biased against Women," *VICE*, October 10, 2018, https://www .vice.com/en/article/amazon-ai-recruitment-hiring-tool-gender-bias.

52 Audrey Amrein-Beardsley, "The Education Value-Added Assessment System (EVAAS) on Trial: A Precedent-Setting Lawsuit with Implications for Policy and Practice," *JEP: eJournal of Education Policy* (Spring 2019), https://eric .ed.gov/?id=EJ1234497; Mark A. Paige and Audrey Amrein-Beardsley, "'Houston, We Have a Lawsuit': A Cautionary Tale for the Implementation of Value-Added Models for High-Stakes Employment Decisions," *Educational Researcher* 49, no. 5 (2020): 350–9, https://doi.org/10.3102/0013189X20923046.

53 Valerie Strauss, "Houston Teachers Sue over Controversial Teacher Evaluation Method," *Washington Post*, April 30, 2014, https://www.washingtonpost.com/news /answer-sheet/wp/2014/04/30/houston-teachers-sue-over-controversial-teacher -evaluation-method/.

54 Amrein-Beardsley, "The Education Value-Added Assessment System (EVAAS) on Trial."

55 During the trial, the proprietor of the algorithm, SAS Institute Inc., allowed one expert witness to access their computer codes and protected information. Despite having far greater access to the system than teachers and the school district, the expert witness was still unable to replicate the scores produced by the algorithm. Audrey Amrein-Beardsley, "The Education Value-Added Assessment System (EVAAS) on Trial"; Audrey Amrein-Beardsley, "VAMboozled!: Houston Lawsuit Update, with Summary of Expert Witnesses' Findings about the EVAAS," National Education Policy Center, January 19, 2016, https://nepc.colorado.edu /es/blog/houston-lawsuit.

56 Warren J. von Eschenbach, "Transparency and the Black Box Problem: Why We Do Not Trust AI," *Philosophy & Technology* 34, no. 4 (2021): 1607–22, https://doi .org/10.1007/s13347-021-00477-0.

57 Yavar Bathaee, "The Artificial Intelligence Black Box and the Failure of Intent and Causation," *Harvard Journal of Law & Technology* 31, no. 2 (2018): 891, https://jolt .law.harvard.edu/assets/articlePDFs/v31/The-Artificial-Intelligence-Black-Box -and-the-Failure-of-Intent-and-Causation-Yavar-Bathaee.pdf.

58 For more reading on this idea, see Andy Clark, *Supersizing the Mind: Embodiment, Action, and Cognitive Extension* (Oxford University Press, 2008).

23. Addressing the Elephant in the Room

1 Carl Benedikt Frey and Michael A. Osborne, "The Future of Employment: How Susceptible Are Jobs to Computerisation?," *Technological Forecasting and Social Change* 114 (January 1, 2017): 254–80, https://doi.org/10.1016/j .techfore.2016.08.019.

2 Songül Tolan et al., "Measuring the Occupational Impact of AI: Tasks, Cognitive Abilities and AI Benchmarks," *Journal of Artificial Intelligence Research* 71 (June 9, 2021): 191–236, https://doi.org/10.1613/jair.1.12647.

3 Chia-Hui Lu, "Artificial Intelligence and Human Jobs," *Macroeconomic Dynamics* 26, no. 5 (July 2022): 1162–201, https://doi.org/10.1017/S1365100520000528; Grace Lordan, "Robots at Work: A Report on Automatable and Non-Automatable Employment Shares in Europe" (Office for Official Publications of the European Communities, 2018), https://eprints.lse.ac.uk/90500/.

4 This idea is not original to me. However, I cannot remember where I originally read it. I thought it was from an Acemoglu article I would have read in 2020, but I cannot find it. If you are reading this chapter and thinking "Hey! I wrote a paper on this!," please email me so I can give you proper credit (and my apologies in advance).

5 Daron Acemoglu and Pascual Restrepo, "Artificial Intelligence, Automation and Work," Working Paper No. 24196, Working Paper Series (National Bureau of Economic Research, January 2018), https://doi.org/10.3386/w24196.

6 Acemoglu and Restrepo, "Artificial Intelligence."

7 Julius Tan Gonzales, "Implications of AI Innovation on Economic Growth: A Panel Data Study," *Journal of Economic Structures* 12, no. 1 (September 9, 2023): 13, https://doi.org/10.1186/s40008-023-00307-w.

8 Timothy F. Bresnahan and M. Trajtenberg, "General Purpose Technologies 'Engines of Growth'?," *Journal of Econometrics* 65, no. 1 (January 1, 1995): 94, https://doi.org/10.1016/0304-4076(94)01598-T.

9 Gonzales, "Implications of AI Innovation on Economic Growth."

10 Yingying Lu and Yixiao Zhou, "A Short Review on the Economics of Artificial Intelligence," SSRN Scholarly Paper (Social Science Research Network, August 7, 2019), https://doi.org/10.2139/ssrn.3433527.

11 Lu and Zhou, "A Short Review."

12 Acemoglu and Restrepo, "Artificial Intelligence," 12.

13 Michael Webb, "The Impact of Artificial Intelligence on the Labor Market," SSRN Scholarly Paper (Social Science Research Network, November 6, 2019), https://doi.org/10.2139/ssrn.3482150; Cecily Josten and Grace Lordan, "Automation and the Changing Nature of Work," *PLOS ONE* 17, no. 5 (May 5, 2022): e0266326, https://doi.org/10.1371/journal.pone.0266326; Tolan et al., "Measuring the Occupational Impact of AI."

14 Pelin Ozgul et al., "High-Skilled Human Workers in Non-Routine Jobs Are Susceptible to AI Automation but Wage Benefits Differ between Occupations" (arXiv, April 9, 2024), https://doi.org/10.48550/arXiv.2404.06472.

15 James Manyika and Kevin Sneader, "AI, Automation, and the Future of Work: Ten Things to Solve For," McKinsey & Company, June 1, 2018, https://www.mckinsey.com/featured-insights/future-of-work/ai-automation-and-the-future-of-work-ten-things-to-solve-for.

16 Acemoglu and Restrepo, "Artificial Intelligence," 12–13.

17 Jeremy Atack, Robert A. Margo, and Paul W. Rhode, "De-skilling: Evidence from Late Nineteenth Century American Manufacturing," *Explorations in Economic History* 91 (January 1, 2024): 101554, https://doi.org/10.1016/j.eeh.2023.101554; Harry J. Holzer, "Understanding the Impact of Automation on Workers, Jobs, and Wages," *Brookings* (blog), January 19, 2022, https://www.brookings.edu/articles/understanding-the-impact-of-automation-on-workers-jobs-and-wages/; Sofia Morandini et al., "The Impact of Artificial Intelligence on Workers' Skills:

Upskilling and Reskilling in Organisations," *Informing Science: The International Journal of an Emerging Transdiscipline* 26 (February 22, 2023): 39–68, https://www.informingscience.org/Publications/5078; Szufang Chuang et al., "Machine Learning and AI Technology-Induced Skill Gaps and Opportunities for Continuous Development of Middle-Skilled Employees," *Journal of Work-Applied Management*, ahead-of-print (November 26, 2024), https://doi.org/10.1108/JWAM-08-2024 -0111; Acemoglu and Restrepo, "Artificial Intelligence," 13.

18 Ignacio González Vázquez et al., eds., *The Changing Nature of Work and Skills in the Digital Age* (Publications Office of the European Union, 2019), https://doi .org/10.2760/679150.

19 David Allan Wolfe and Tracey M. White, "Canada as a Learning Economy: Education and Training in an Age of Machines – Policy Challenges and Policy Responses," *Skills and Work in the Digital Economy Conference*, Ottawa, ON, September 2021, https://utoronto.scholaris.ca/items/b956046b-3d33-4d3a-bf68 -d4631f6519a1.

20 Tom Lehman, "Countering the Modern Luddite Impulse," *The Independent* 20, no. 2 (2015): 265–83, https://www.independent.org/wp-content/uploads /tir/2015/10/tir_20_02_05_lehman.pdf.

21 For further reading on the Industrial Revolution and the Luddites, see Katharine McGowan and Sean Geobey, "'Harmful to the Commonality': The Luddites, the Distributional Effects of Systems Change and the Challenge of Building a Just Society," *Social Enterprise Journal* 18, no. 2 (February 9, 2022): 306–20, https:// doi.org/10.1108/SEJ-11-2020-0118; David Taylor, "The Luddites," in *Mastering Economic and Social History* (Macmillan Education, 1988), 71–9, https://doi .org/10.1007/978-1-349-19377-6_5; Katrina Navickas, "The Search for 'General Ludd': The Mythology of Luddism," *Social History* 30, no. 3 (August 1, 2005): 281–95, https://doi.org/10.1080/03071020500185406; Richard Conniff, "What the Luddites Really Fought Against," *Smithsonian Magazine*, March 2011, https://www .smithsonianmag.com/history/what-the-luddites-really-fought-against-264412/; John Foster, *Class Struggle and the Industrial Revolution: Early Industrial Capitalism in Three English Towns* (Routledge, 2003).

22 E.J. Hobsbawm and Chris Wrigley, *Industry and Empire: From 1750 to the Present Day* (Penguin, 1999); David S. Landes, *The Unbound Prometheus: Technological Change and Industrial Development in Western Europe from 1750 to the Present*, 2nd ed. (Cambridge University Press, 2003); Robert C. Allen, *The Industrial Revolution: A Very Short Introduction* (Oxford University Press, 2017).

Index